DOGS

DOGS

THEIR FOSSIL RELATIVES AND EVOLUTIONARY HISTORY

XIAOMING WANG + RICHARD H. TEDFORD | ILLUSTRATIONS BY MAURICIO ANTÓN

COLUMBIA UNIVERSITY PRESS   NEW YORK

COLUMBIA UNIVERSITY PRESS
Publishers Since 1893
New York   Chichester, West Sussex

Library of Congress Cataloging-in-Publication Data
Wang, Xiaoming, 1957–
    Dogs : their fossil relatives and evolutionary history / Xiaoming Wang and
    Richard H. Tedford ; illustrations by Mauricio Antón.
        p. cm.
    Includes bibliographical references and index.
    ISBN 978-0-231-13528-3 (cloth) — ISBN 978-0-231-13529-0 (pbk.) — ISBN
978-0-231-50943-5 (e-book)
    1. Dogs. 2. Canis, Fossil. 3. Dogs—Evolution.
    I. Tedford, Richard H. II. Antón, Mauricio. III. Title.

QL737.C22W36   2008
599.77'2—DC22
                                                                      2007052328

BOOK DESIGN *by* VIN DANG + JACKET DESIGN *by* MARTIN HINZE

CONTENTS

PREFACE

THE LONG ASSOCIATION OF HUMANS and the domestic dog has attracted considerable interest in the biology and natural history of the Canidae (dog family), among both specialists and the general public. Several recent events have converged to inspire further interest in canids. New discoveries of early domestic dogs living in Israel and Germany more than 14,000 years ago push back the fossil record and provide tantalizing evidence of initial interactions between early humans and their first domestic animal. Meanwhile, molecular studies indicate that China, not Europe or the Middle East, may have been the center of first domestication. Moreover, humans' fascination with large predators such as wolves and hunting dogs has continued to draw public attention to a variety of issues ranging from predatory behavior to ecology and conservation. The conservation and reintroduction of gray wolves in Europe and North America and of red wolves in the southeastern United States have promoted public debate regarding the need to protect large predators in order to achieve a balanced ecological community.

Against this background of sustained public interest in everything about dogs, the unparalleled Frick Collection (American Museum of Natural History) of North American Cenozoic canids has been the subject of three monographs by Xiaoming Wang (curator of Vertebrate Paleontology), Richard Tedford (curator emeritus of Vertebrate Paleontology), and Beryl Taylor (curator of the Frick Collection) (Wang 1994; Wang, Tedford, and Taylor 1999; Tedford, Wang, and Taylor in press). The wealth of information provided in these monographs reveals a vast array of extinct canids in North America, the continent of their origin, and their subsequent expansion to the Old World about 5 to 7 million years ago (Ma) and to South America around 3 Ma. These immigration events ultimately helped to make canids one of the most widely spread living carnivores in the world and the dominant predators on some continents. This book attempts to capture the excitement of recent advances in our understanding of the natural history of this family of carnivorans. The book is richly illustrated, both to provide an unprecedented visual reference for the specialists and to make it accessible and interesting to a wide nontechnical audience.

The family Canidae was the earliest group that still has extant members to arise within the order Carnivora. During their history of more than 40 million years, canids have also achieved a longevity and diversity unrivaled by any other group of carnivores. Their success is also exemplified by a worldwide distribution and achievement of top predator status in modern North and South America, Australia, and northern Eurasia—only in Africa are they less dominant among large predators. It is argued that the propensity for canids to form large, social hunting groups and the associated development of their brain were crucially important in the domestication process—humans learned to form a mutually beneficial relationship with such an intelligent carnivoran that was preadapted to life among a different species. The importance of the domestication of various animals in early human societies cannot be overestimated, and it may have been first inspired by the domestication of dogs.

The evolutionary history of canids is a history of successive radiations (rapid expansions of diversity within a group of organisms, often in response to environmental changes or new resources) repeatedly occupying a broad spectrum of niches, ranging from large, pursuit predators to small omnivores and possibly even to herbivores. Three such radiations were first recognized by Richard Tedford (1978), each represented by a distinct subfamily. Two archaic subfamilies, Hesperocyoninae and Borophaginae, thrived in the middle to late Cenozoic from about 40 to 2 Ma. All extant canids belong to the final radiation, the subfamily Caninae, which achieved its present diversity only in the past few million years.

Previous diversities of canids were intimately related to environmental changes during the past 40 million years. Large, wolflike pursuit predators were in particular the direct consequence of the opening up of the landscape. A worldwide change from a warm and humid climate to an increasingly colder and drier climate resulted in large-scale transformation of forested environments to open grasslands worldwide. These changes were reflected in the terrestrial vertebrate community by an increasing cursoriality (the ability to run fast and for a long time) among predators and prey. This greater ability to run is clearly reflected in the fossil history of canids, which shows increasingly erect standing postures, progressively lengthened limb bones, and restricted movements of the joints.

Humans' fascination with canids is derived from two main sources: our love affair with our dogs and our reverence of their top predator status in much of the world. Such a high interest in a group of mammals is generally rare and offers a great opportunity to engage the public in paleontology, functional morphology, evolution, and behavioral ecology.

ACKNOWLEDGMENTS

WE THANK OUR PUBLISHERS at Columbia University Press, Patrick Fitzgerald and Robin Smith, and editor Irene Pavitt for their encouragement and helpful guidance. We are indebted to Lars Werdelin and Blaire Van Valkenburgh for their critical reviews, which brought to our attention records that we missed and other factual errors. A thorough editing by copyeditor Annie Barva has greatly improved the language and expression of this book. Alejandra Lora of the American Museum of Natural History typed part of the manuscript for publication and helped in numerous other ways to facilitate communications. A substantial part of the monographic treatments of the three subfamilies of canids were funded by two grants from the National Science Foundation (DEB-9420004 and DEB-9707555) and a Frick Postdoctoral Fellowship from the American Museum of Natural History. Xiaoming Wang thanks his wife, Yanping, and his son, Alex, for their support and understanding throughout the preparation of this book. Richard Tedford is grateful for the patience of Elizabeth and Vivien during the 30 years this study took to complete. Mauricio Antón thanks his wife, Puri, and his son, Miguel, for their patience during this long process.

1 METHODS OF STUDY AND THE PLACE OF DOGS IN NATURE

THE BASIS FOR A CONSISTENT CLASSIFICATION of living organisms began in the mid-eighteenth century with Carolus Linnaeus's (Karl von Linné) monumental work *Systema naturae*. In the tenth edition of that work in 1785, he proposed that the group *Canis* include three genera—*Canis*, *Vulpes*, and *Hyaena*—based on similarities in form and function (such as wolflike and foxlike predations). The hyenas were later placed in their own family, Hyaenidae, within the order Carnivora, and the family Canidae, also within that order, was to hold *Canis*, represented by the wolf (*C. lupus*), the fox (*Vulpes vulpes*), and a number of other species formalized during the nineteenth century (Mivart 1890). In this book, we use the term *dog* informally to denote members of the family Canidae rather than just the domestic dog (*Canis familiaris* or *Canis lupus familiaris*).

It was recognized early on that all members of the order Carnivora (collectively called *carnivorans*) possess a common arrangement of teeth, in which the last upper premolar and lower first molar have bladelike enamel crowns that function together as shears (figure 1.1). This dental arrangement was a fundamental adaptation to cutting meat, and all carnivorans are characterized by the possession of these teeth, which are called *carnassials* (chapter 4). Through the long history of the Carnivora (more than 60 million years), this dentition has been modified for many adaptations—some focused exclusively on a diet of meat (hypercarnivory [chapter 4]), some on the crushing of vegetation (hypocarnivory), and some on generalized ("primitive") functions (mesocarnivory) or the loss of the carnassial function altogether, as in the termite-eating hyaenid *Proteles*, fish-catching seals and sea lions, and mollusk-crushing walruses. This diversity of dental adaptations provides many ways to characterize species and some species groups (genera or families), but the form of the carnassials themselves limits the number of ways that they can function. In the past, the chance of similar adaptations unrelated to genealogy was great and

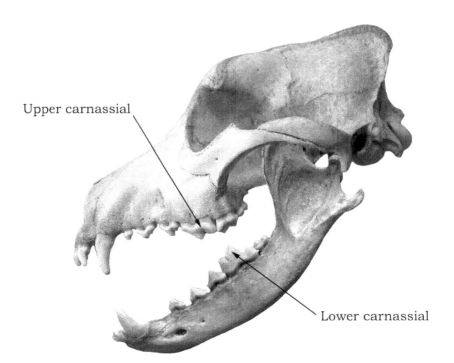

Upper carnassial

Lower carnassial

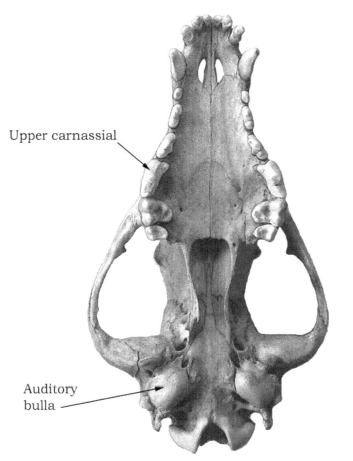

Upper carnassial

Auditory
bulla

could confuse a classification intended to give evidence of evolutionary relationships. This was not a problem for Linnaeus or the eighteenth-century biologists, but it did become a concern for Charles Darwin and other nineteenth-century evolutionary biologists who followed him.

The British anatomist William Henry Flower (1869) noticed that the form and composition of the bony structures that surround the middle ear of the carnivoran skull provide another means of subdividing the Carnivora into three groups based on features other than the teeth. These structures include the size, form, and composition of the bony covering at the base of the skull that encloses the middle-ear space in which lay the tiny auditory ossicles (bones) essential to hearing (see figure 1.1). This bony structure, called the auditory bulla, is formed of two major sites of ossification. The ectotympanic bone contains the external opening of the middle-ear space and incorporates the bony tympanic ring that holds the tympanic membrane ("eardrum"), into which the malleus, the outermost part of the ossicular chain, is inserted. The entotympanic bone forms the inner and most inflated part of the bulla, often enclosing part of the internal carotid artery. These two parts of the bulla are fused together, the junction often prolonged internally to form a septum dividing the middle-ear cavity into two chambers. Flower found that the living carnivoran families form three large groups according to the shape and construction of the bulla. The catlike carnivores (Aeluroidea) have an inflated bulla with a large septum that nearly subdivides the middle-ear space (see figure 4.10). This septum is derived in nearly equal contributions from the ectotympanic bone and the entotympanic bone, but the entotympanic encloses the internal carotid artery. The bearlike carnivores (Arctoidea), in contrast, have a little inflated bulla that lacks a septum, but the internal carotid artery is enclosed by the entotympanic bone. The doglike carnivores (exclusively Canidae and Cynoidea) have an inflated bulla and a low, unilaminar septum composed largely of the entotympanic bone, but the internal carotid is enclosed for only a short distance anteriorly in the latter bone. Once identified, these bony features and the distinctions among them in different carnivoran families became of great value to paleontologists in their attempts to group carnassial-bearing fossils into evolutionary units.

With Charles Darwin's introduction in 1859 of natural selection as the mechanism of evolution, evolution became more widely accepted. The fossil record of animals and plants had also been accepted as evidence that a history of life was contained

FIGURE 1.1

Skull and mandible of a domestic dog

Top, lateral view; *bottom,* ventral view.

in the geological record. This acceptance enabled evolutionary reconstructions to be carried into the geological past. Early evolutionists were unaware of the genetic bases for evolution until Gregor Mendel's work was discovered in the early twentieth century, but discovery of the means of genetic change had to wait until later in that century.

Most twentieth-century biologists and paleontologists continued to describe the form, or *morphology*, of canids in order to use resemblance as a tool to establish biological relationships consistent with evolution. Because species are the smallest units usually dealt with, the nature of these units needed definition. At midcentury, with the acceptance of genetic and morphological bases for definition, a species was regarded either as a *biological species* (a group of actually or potentially interbreeding natural populations that are reproductively isolated from other such groups [Mayr 1940]) or as a *morphological species* (a group of individuals or populations with the same or similar morphological characters). The range of morphological variation among the individuals recognized as a species would approximate those characters observed in related living species. The morphological definition is the only practical approach when considering the fossil record and living species where interbreeding in nature has not been observed.

The determination and definition of canid species are thus fundamental to our quest for the history of the family Canidae, or its *phylogeny*. We have used the principles of phylogenetic systematics in determining evolutionary relationships among species because we believe this approach leads to conclusions that can be tested by the hypothetical-deductive methods most consistent with modern science. Hypotheses about the evolutionary relationships between species are demonstrated by the common possession of specific features unique to these species and derived from a single source common to both (*homology*). This evidence indicates a *monophyletic* relationship between the species if the ancestor in common is not found in common with any other species. Such statements are dependent on the quantity of the evidence and are intrinsically better tested among living species than among fossil species, where the evidence is restricted not only to the skeleton, but also to its state of preservation.

This kind of phyletic analysis is often termed *cladistic* because it reconstructs *clades*, or monophyletic branches, into a "cladogram," or diagram, of mutual relationships based on a specific set of morphological features. Inadequate evidence may lead to several alternative fits of the data underpinning the cladogram. In these cases, the principle of parsimony guides the first choice among fits: the simplest explanation of the cladogram is most likely the best.

In this book, we discuss the evolution of the Canidae. We include some features that are not demonstrably cladistic, but that seem compatible with more general

AGE (Ma)	Epoch	Period	Era
	Pleistocene		
5	Pliocene	Neogene	
10	Miocene		
15			
20			
25	Oligocene		Cenozoic
30			
35			
40	Eocene	Paleogene	
45			
50			
55	Paleocene		
60			
65	Cretaceous		Mesozoic

FIGURE 1.2

Time scale of the Cenozoic era

biological relationships, in particular those that suggest direct or anagenetic processes in evolution rather than bifurcating, cladogenetic ones. We also place the phylogenies in time so that the tempo of biological change is shown and thus the history of the canid family can be depicted (appendix 2).

The time scale for the history of canids, most conveniently expressed in units of millions of years ago (figure 1.2), has been developed from more than 200 years of study, first only of the rock record (stratigraphy) in which the natural superposition of strata (younger over older) provides evidence of the relative ages of the contained fossil samples, but later from correlation using a variety of tools such as specific paleontological similarity and radioactive elemental isotopes that give direct ages of rocks based on known rates of elemental decay.

2 THE ORIGIN OF CANIDS AND OTHER DOGLIKE CARNIVOROUS MAMMALS

DOGLIKE PREDATORS, WHETHER TRUE CANIDS OR NOT, have always played an important role in the predatory community. Therefore, understanding the competitive landscape requires knowing about the other carnivorous mammals that existed before and during the emergence of the canids as well as about the doglike carnivores that lived on the continents before canids arrived.

The World of Carnivorous Mammals

The order Carnivora (from Latin *carnis* [flesh] and *vorare* [to devour]) includes all mammals that possess a pair of shearing carnassial teeth formed by the upper fourth premolar and lower first molar. All members of this order descended from an ancestor that possessed this character, and they form a natural group because of their common ancestry. Throughout this book, we refer to members of this group as *carnivorans* instead of *carnivores*, which covers a broader array of predators—such as mesonychids, creodonts, borhyaenids, and thylacinids—that do not possess this defining feature, the carnassial teeth. It is important to keep in mind that the latter carnivorous mammals (predators that consume meat as their main diet) do not form a natural group because they are not closely related to one another. The only attribute they have in common is their preference for meat, which by itself does not constitute a valid criterion for distinguishing them as a natural group (for example, numerous reptiles—such as theropod dinosaurs, crocodiles, and snakes—are also carnivores, but each has its own separate origin and natural group).

A few groups of carnivores were important predators prior to the origin of canids or were contemporaneous canidlike carnivores that either directly competed with canids or occupied similar niches in places where canids were absent. Canidlike predatory adaptations were common throughout the Cenozoic era (the past 65 million years [see figure 1.2]), so it can be concluded that the canid body plan was and continues to be a very successful design for chasing and subduing prey.

CIMOLESTIDS

Ancestral mammals during the Mesozoic era (248 to 65 Ma) lived under the shadow of the dinosaurs. They were largely mouse- or rat-size animals similar to today's insectivores (hedgehogs, shrews, and so on), although an occasional dog-size mammal could hunt for small dinosaurs (Hu et al. 2005). During the late Cretaceous (75 to 65 Ma) in North America, a group of rat-size mammals called cimolestids began to develop progressively more bladelike cheek teeth apparently adapted for cutting flesh. In the species *Cimolestes cerberoides* from the Scollard Formation of Alberta, Canada, the upper fourth premolar began to assume the form of an upper carnassial tooth with the development of a shearing crest formed by the paracone and metacone—the tooth that was to become typical of carnivorans. Such a shearing crest at exactly the same location as for all later carnivorans offers tantalizing evidence that carnivorans' ancestry may be traced to the late Mesozoic (75 Ma), even though the small body size and limited dental specialization suggests that *Cimolestes* probably was not an exclusively meat eater.

VIVERRAVIDS AND MIACIDS

True carnivorans, members of the order Carnivora, arose shortly after the Cretaceous–Tertiary mass extinction around 65 Ma. The first carnivoran to appear in the early Paleocene (65 to 60 Ma) of North America belongs to the family Viverravidae, an extinct family that featured a true pair of carnassial teeth. Shortly after this first appearance, the viverravids apparently spread first to Asia and later to Europe, although records of this family are so poor that much remains to be learned about it. The viverravids possessed a reduced set of dentition (loss of the last, or third, molars) from the very beginning, and this curious feature has been used to indicate relationships with the catlike carnivorans (Feliformia), which include families such as the Felidae and the Hyaenidae, because the feliforms also have a reduced molar dentition. However, this connection is weakened by a lack of transitional fossil records in the Eocene (55 to 35 Ma) that bridge the gap between the earliest progenitors of various feliform families and the viverravids. Recent studies by Gina D. Wesley-Hunt and John J. Flynn (2005) attempt to reconcile this contradiction, postulating that the viverravids were an early specialized carnivoran group unrelated to any living families.

Another group of archaic carnivorans is the family Miacidae, which first appeared in the late Paleocene to early Eocene (60 to 50 Ma) of North America and

Europe and later spread to Asia. Like the viverravids, the first miacids also featured a true pair of carnassial teeth, signaling their relationship to the order Carnivora. In contrast to viverravids, however, the miacids were more generalized in their dental adaptations because of the primitive possession of a full complement of cheek teeth (presence of upper and lower third molars) (figure 2.1). Miacids were weasel-size to small fox–size predators (and occasionally small dog–size species) living in forested terrain and, like the viverravids, were limited to relatively small prey (figure 2.2). The true significance of the miacids thus lies not in their ecological diversity and impact on the prey community, but in their ancestral relationship to later carnivorans. From various lineages within the miacids, a number of, or possibly all, modern families of Carnivora arose.

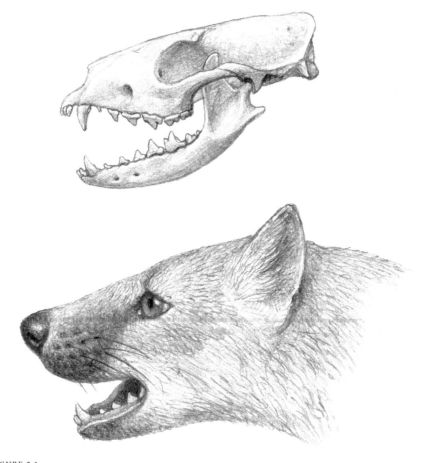

FIGURE 2.1

Miacis sylvestris

Skull and reconstructed head of *Miacis sylvestris*, based on fossils from the Eocene Wind River Formation (50 Ma) of Wyoming. Approximate length of skull: 10 cm.

FIGURE 2.2

Miacis kessleri

Life reconstruction of the miacid *Miacis kessleri*, based on an articulated skeleton from Messel in Germany (50 to 36 Ma).

AMPHICYONIDS

Another extinct group of doglike carnivorans is the family Amphicyonidae. This great family of "bear dogs," as they are popularly known, was very important within the predatory community of the past. As the common name implies, the bear dogs look like an intermediate form between bears and dogs. The amphicyonids are so similar to the canids that for many years paleontologists had difficulty distinguishing the two families. Aside from a third upper molar in the early forms, the amphicyonids' teeth share numerous similarities with canids' teeth, including the carnassials that mark them as true carnivorans. In their primitive dental formula and the tendency toward reduction in the size of the premolars, the amphicyonids also resemble the ursids. Amphicyonids also tend to have a canidlike skull and body proportion that includes an elongated rostrum and long leg bones. Early forms such as *Daphoenodon* (figure 2.3), from the earliest Miocene (23 Ma) of North America, had elongated feet and a digitigrade stance (they walked on the digits, with the posterior of the foot raised), but the hands and feet of the later, larger genera became semiplantigrade (they walked on the sole, with the heel touching the ground) (Hunt 2003), tending to resemble bears.

Given these morphological ambiguities, it is not surprising that amphicyonids have been a subject of controversy for some time, further exacerbated by the fact that they are long extinct, depriving us of the soft-anatomical information that would help us to understand the relationship between them and the canids. Some early paleontologists went as far as placing bear dogs within the family Canidae. However, more detailed examinations of the bear dogs' ear region, most recently by Robert M. Hunt Jr. (2003), have revealed that amphicyonids are probably more closely related to the bear family, Ursidae (this is not to say that amphicyonids are bears, but that amphicyonids and ursids are technically sister groups).

Amphicyonids first appeared in the middle Eocene (45 Ma) of North America and quickly spread to Europe in the late Eocene (35 Ma) and to Asia and Africa by the early Miocene (23 Ma). By the late Miocene (8 Ma), amphicyonids became extinct on all continents. Therefore, amphicyonids were contemporaneous with North American hesperocyonine and borophagine canids. In fact, many of the North American amphicyonids were of a size similar to that of contemporary canids; this considerable overlap in body size as well as their cranial and dental adaptations indicates that they must have competed with canids. Interestingly, a number of amphicyonid lineages also independently evolved dental modifications toward hypercarnivory (specializing in meat eating [chapter 4]) and a progressively larger body size, so their clash with canids over prey resources was probably inevitable. In such a competitive landscape, only one feature stands out for canids—its digitigrade posture. Although it is not possible to be sure that a more erect posture afforded canids a decisive advantage, the fact remains that canids outlasted amphicyonids; when amphicyonids were approaching extinction in the late Miocene, canids were just beginning to spread to the rest of the world.

FIGURE 2.3

Daphoenodon superbus

Life reconstruction of the amphicyonid *Daphoenodon superbus*, from the early Miocene (23 Ma) of Nebraska. Unlike later, larger members of the family Amphicyonidae, *Daphoenodon* had rather elongated feet and a digitigrade stance. Reconstructed shoulder height: 59 cm.

FIGURE 2.4

Ictitherium ebu

Life reconstruction of the hyaenid *Ictitherium ebu*, from the late Miocene (6 Ma) of Lothagam, Kenya. Reconstructed shoulder height: 60 cm.

HYAENIDS

Although phylogenetically not closely related to canids, living hyenas (family Hyaenidae), with the exception of the highly specialized aardwolf (which is a peculiar termite specialist), share many similarities with canids. Both families are highly cursorial (adapted to running) and are well suited for pursuit in open grasslands. Members of each of these families also form elaborate hunting packs and have complex social behaviors. Furthermore, both families include species that are adapted to bone crushing. In fact, the modern spotted hyena (*Crocuta crocuta*) is the best living analogue for the bone-cracking borophagine canids (chapter 3).

The history of hyaenid evolution also resembles that of canid evolution in a number of ways. As North America is the continent of origin for canids, Eurasia is the continent of origin for hyaenids. Each of these families essentially stayed within the continent of its origin for much of its early existence. Canids eventually reached out to the rest of the world during the late Miocene and into the Pliocene (5 to 1.8 Ma), whereas hyaenids almost never made it to the New World, with the single exception of *Chasmaporthetes*, which briefly expanded to North America during the Pliocene (chapters 4 and 5).

Hyaenids also started out as fox-size forms in the early to middle Miocene, such as *Protictitherium*, *Tungurictis*, and *Plioviverrops*. Through transitional forms such as the coyote-size *Ictitherium* (figure 2.4) and the wolf-size *Hyaenictitherium* (figure 2.5), hyaenids evolved to become powerful bone crushers, such as *Pachycrocuta*, *Adcrocuta*, and *Crocuta*. Some lineages appear to have been more adapted for running in open terrain, such as *Chasmaporthetes*, which is the only genus that managed to cross Beringia (a periodic land bridge between eastern Siberia and Alaska, now the Bering Strait) to North America.

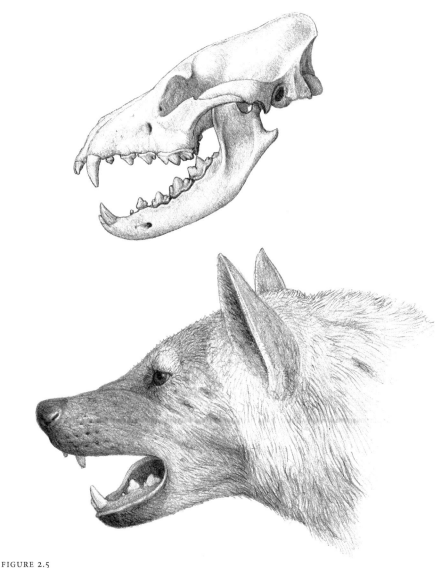

FIGURE 2.5

Hyaenictitherium wongi

Skull and reconstructed head of the hyaenid *Hyaenictitherium wongi*, from the late Miocene (8 Ma) of China. Total length of skull: 18 cm.

FIGURE 2.6
Synoplotherium vorax
Life reconstruction of the mesonychid *Synoplotherium vorax*, based on fossils from the Eocene Bridger For-
mation (45 Ma) of Wyoming. Reconstructed shoulder height: 58 cm.

MESONYCHIDS

The viverravids and miacids were predominantly cat- or fox-size predators, unable
to handle the larger prey of the Paleocene and Eocene. To fill such a void, a group
of primitive even-toed ungulates (artiodactyls, hoofed mammals including the forerunners of hippos and cows) evolved in the northern continents (Europe, Asia, and
North America) and played the role of large terrestrial predators. The mesonychids, such as *Synoplotherium* (figure 2.6) and *Sinonyx* (figure 2.7), were superficially
wolflike mammals that often were the top predators of their day. They had strong
canines and robust cheek teeth and were probably capable of crushing bones. Although their long legs suggest the capability of catching fast prey, some paleontologists thought they were scavengers because of their strong teeth. Perhaps the most
famous member of the mesonychids is the giant *Andrewsarchus mongoliensis* from
the middle Eocene of Central Asia, probably the largest mammalian predator on
land at the time. Named after Roy Chapman Andrews, leader of the Central Asiatic
Expeditions of the American Museum of Natural History, *Andrewsarchus* has an
enormous skull of nearly 1 meter long and massive teeth capable of crushing bones.

CREODONTS

Although some mesonychids may have been the top predators in some parts of the
world because of their sheer size and robust jaws, the truly dominant predators in
the Eocene were the creodonts (order Creodonta). Among the creodonts, members
of the family Hyaenodontidae were the most specialized predators of their time.

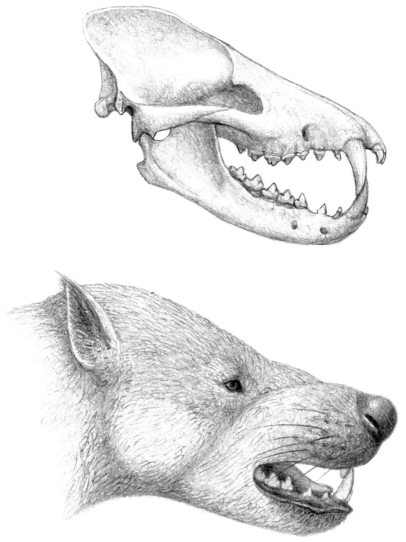

FIGURE 2.7

Sinonyx

Skull and reconstructed head of the mesonychid *Sinonyx*, from the Eocene (around 50 Ma) of China. Total length of skull: 32 cm.

Instead of one pair of carnassial teeth, as in the Carnivora, hyaenodonts sported as many as three sequential pairs of carnassial-like molar teeth in their jaws. These posteriorly located cheek teeth were kept sharp all the time and were superb devices for cutting meat and tendons. Hyaenodonts had elongated skulls, high sagittal crests for attachment of powerful temporalis muscles for jaw occlusion, and long limbs,

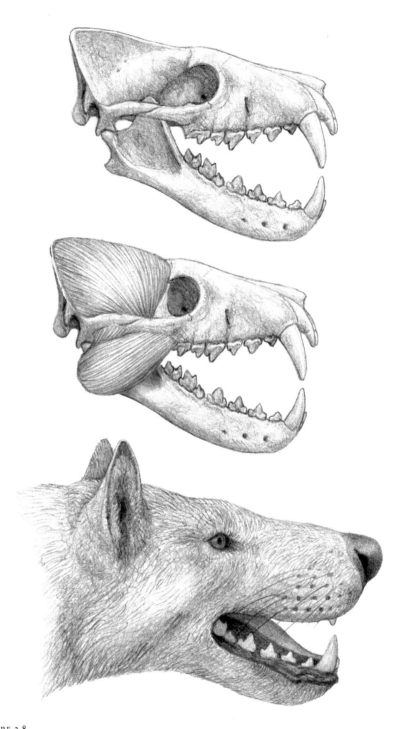

FIGURE 2.8

Hyaenodon horridus

Skull, masticatory musculature, and reconstructed life appearance of the head of *Hyaenodon horridus*. Total length of skull: 35 cm.

FIGURE 2.9

Comparison of *Hyaenodon horridus* and *Hyainailouros sulzeri*

Life reconstructions of two hyaenodontid creodonts, shown to the same scale. *Foreground, Hyaenodon horridus*, from the early Oligocene (33 Ma) of North America; *background, Hyainailouros sulzeri*, from the early Miocene (20 Ma) of France. Hyaenodontids had a vaguely doglike appearance, but their heads were very large in proportion to their bodies. Reconstructed shoulder height of *Hyainailouros*: 1 m.

features often associated with wolflike adaptations (figures 2.8 and 2.9). Hyaenodonts, however, lacked the dental versatility of canids (that is, the presence of postcarnassial crushing molars), and their teeth were not suited for processing a variety of foods other than meat. *Hyainailouros*, one late lineage of the hyaenodonts in the early to middle Miocene (22 to 14 Ma) of Europe, Africa, and southern Asia, eventually evolved to a giant size and probably developed bone-crushing adaptations. Although not quite as extremely adapted for shearing meat as *Hyaenodon, Hyainailouros* certainly rivaled the contemporaneous *Hyaenodon* in size and strength, vying for top-predator status of their time.

Hyaenodonts thrived in the Eocene of Eurasia and North America and continued to play an important role in the predatory community well into the Miocene of Africa and southern Asia. Although finally outcompeted by true carnivorans on all continents, hyaenodonts held their own and probably played a significant role in shaping the predatory landscape that profoundly influenced the early evolution of many families of carnivorans.

BORHYAENIDS

Through much of the Cenozoic, South America was essentially an island continent separated from its northern neighbor, North America, by a seaway (for the Great American Biotic Interchange during the Pliocene, see chapters 6 and 7). In near total isolation from the rest of the world, marsupial mammals in South America

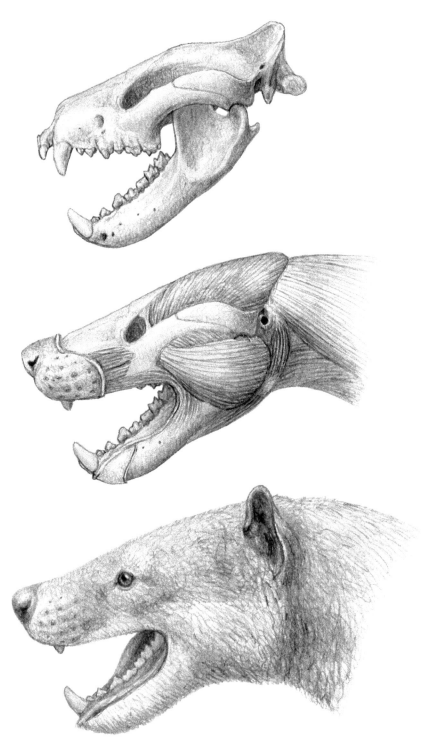

FIGURE 2.10

Borhyaena tuberata

Skull, musculature, and reconstructed head of the South American marsupial *Borhyaena tuberata*. Some of this animal's premolars were enlarged and adapted for bone crushing; they vaguely resemble the teeth of a hyena, a similarity indicated in this species's Latin name. Total length of skull: 23 cm.

FIGURE 2.11

Borhyaena tuberata

Life reconstruction of the South American marsupial carnivore *Borhyaena tuberata*. Although vaguely doglike in appearance, this animal was short-legged with a rather large head. Reconstructed shoulder height: 36 cm.

had gone their own independent ways. In predatory communities, members of the family Borhyaenidae became the de facto top predators. Without competition from the rest of the world, the borhyaenids quickly diversified in the early Cenozoic and included a variety of predators from the Paleocene to the Pliocene. Some of these borhyaenids, such as *Borhyaena* in the early Miocene (22 Ma) of Argentina, were canidlike predators (figures 2.10 and 2.11). *Borhyaena* has a large head, a long neck, and relatively short legs. These characteristics suggest that *Borhyaena*, like the wolf, was probably adapted to a variety of habitats.

The borhyaenids were the dominate mammalian predators throughout the Cenozoic of South America, but had to give up that status when true placental carnivorans came across the newly formed Isthmus of Panama in the middle Pliocene (about 3 Ma). Canids, as the most diverse modern carnivorans in South America, presumably played a major role in the demise of the borhyaenids (chapter 7).

THYLACINIDS

Compared with South America, Australia was even more isolated as an island continent beginning in the middle Cenozoic. With the exception of bats, rodents, and egg-laying monotremes, Australian marsupials made up the majority of the mammalian community until the dingo arrived about 5,000 years ago (chapter 8). Among indigenous marsupial predators, the fabled Tasmanian wolf (*Thylacinus cynocephalus*) was perhaps the closest to a doglike carnivore in Australia. Extinct in the Australian mainland before the Europeans arrived, the Tasmanian wolf survived on the island of Tasmania, only to be hunted to extinction in the twentieth century. Its body plan and skull shape is quite canidlike, with a lengthened rostrum and long

FIGURE 2.12

Nimbacinus dicksoni

Reconstructed life appearance of the marsupial *Nimbacinus dicksoni*, from the Miocene (22 Ma) of Riversleigh, Australia. Smaller and less specialized for carnivory than its relative, the recently extinct Tasmanian "wolf," this Miocene animal could more appropriately be termed a marsupial "fox." Reconstructed shoulder height: 30 cm.

legs. However, many detailed differences in the skull and teeth leave no doubt that the thylacinids are not canids. The superficial similarities between the Tasmanian "wolf" and true canids are good examples of convergent evolution, in which functionally similar structures evolved in groups that have very distant genealogical relationships. The thylacinids can be traced to the late Oligocene (25 Ma) of Australia, and the early forms such as the Miocene species *Nimbacinus dicksoni* (figures 2.12 and 2.13) were smaller and less-specialized predators than the Tasmanian wolf.

The Origin of Canids and Their Fossil Record

The family Canidae had its origin in North America and spent most of its history on that continent. Carnivora in general are usually rare and fragmentary in fossil collections, but canid remains are relatively frequent and often complete enough to offer significant comparative material for phyletic analysis. Because of this fortunate circumstance, the phylogeny of canids can be reconstructed in more detail than can the phylogeny of most carnivorans, and fewer canid clades are missing important steps in their histories.

Some regions of North America have been particularly informative in tracing the canids' history. Deposits in the Great Plains of the United States extending east from the Rocky Mountains across Texas, Colorado, Nebraska, Wyoming, and South Dakota have produced long geological records from the late Eocene (42 Ma) to the late Pleistocene (0.01 Ma) in a discontinuous fashion with intervening geological

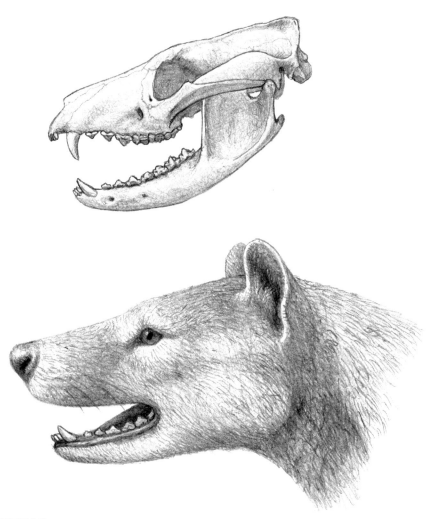

FIGURE 2.13
Nimbacinus dicksoni
Skull and reconstructed head of the marsupial *Nimbacinus dicksoni*. Total length of skull: 13 cm.

hiatuses, but with sufficient continuity to show the average rates of canid evolution and thus usefully to establish much of the group's history. This region continues to harbor the records of two-thirds of the native canid species of North America. Other important records that depict shorter spans of time are scattered across western North America: the early Cenozoic basins within the Rocky Mountains give evidence of the origin of the family; important records of Miocene history are found in rocks in the Great Basin of California (Mojave Desert), the Rio Grande in New Mexico (Albuquerque and Española basins), the northern Rockies of Montana, and the Gulf coasts along Texas and Florida. These records give geographic breadth to

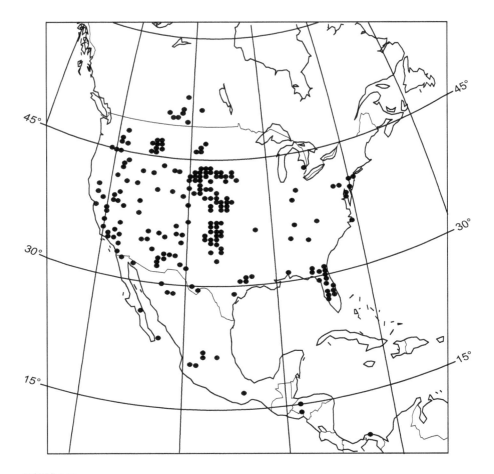

FIGURE 2.14

Distribution of canid sites in North America

Known fossil localities are concentrated in midlatitudes (30° to 40° North) in the western states of the United States, where Cenozoic deposits are best exposed and most intensively explored. (Map and data points supplied by the Paleobiology Database at http://paleodb.org)

the evidence of the origin of specific groups of canids, some first appearing in specific regions before spreading out more widely (figure 2.14).

With the union of North and South America in the late Cenozoic (3 Ma) and the previous high-latitude junction with Eurasia (around 7 Ma), canids made their first move outside North America. Products of canid evolution on these continents, especially in Eurasia, later began to return to North America and to interact with the native canid fauna. The high-latitude canid fauna, including the wolf and the red and arctic foxes, is of Eurasian origin. These connections resulted in family members achieving intercontinental ranges and stimulating widespread evolution of canid clades adapted to new physical and biological environments.

3 DIVERSITY
WHO IS WHO IN THE DOG FAMILY

THERE HAVE BEEN MORE THAN 214 species of canids throughout the Canidae's 40-million-year history (appendix 1). Excluding the 37 species that are still with us today, 177 extinct species are known so far in the fossil records of the world, although the number will surely grow as discoveries of new fossil species are made around the globe. In this chapter, we briefly describe a few species from each subfamily to give a sense of the diverse world of canid evolution.

Hesperocyonines

The subfamily Hesperocyoninae is named after the genus *Hesperocyon*, a small, primitive canid that gave rise to most of the later canids. The hesperocyonines are the most ancient subfamily of the canids. They were the earliest to arise, dating back to almost 40 Ma during the later part of Eocene time, and they were also the first to become extinct. Their last member, *Osbornodon fricki*, lived around 15 Ma, in the middle Miocene. The subfamily's history in North America thus spans approximately 25 million years (appendix 2). Fossil remains of the hesperocyonines have been found over much of the western United States, and a few have been found in Canada and Mexico.

The hesperocyonines evolved from an archaic group of carnivorans in the family Miacidae. In particular, a series of fox-size species within the genus *Miacis* may have progressively given rise to the first definitive canid, *Prohesperocyon*. The first sign that this small carnivore had become a canid can be found in its ear region, where a protective bony housing, the auditory bulla, developed to shield the extremely delicate middle-ear bones (also known as middle-ear ossicles, which include the malleus, incus, and stapes) (figure 3.1). All hesperocyonines also have a moderately sharp, wedge-shaped lower first molar (the lower carnassial [chapter 4]) and narrow upper molars, features that help paleontologists to recognize canids.

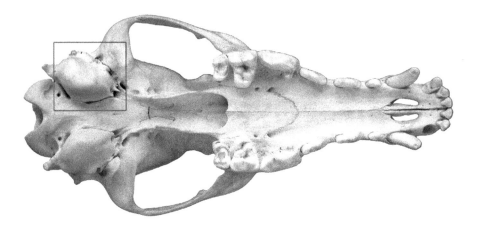

FIGURE 3.1

Auditory bulla in a canid skull

In this coyote (*Canis latrans*) skull can be seen the structure of the bulla (within the square), a bony housing that protects the tiny ear bones (malleus, incus, and stapes) that link the eardrum and the inner-ear cavity and serve the important function of sound transmission (see also figures 4.10 and 4.11).

By the end of the Eocene (37 Ma), species of the genus *Hesperocyon* began to appear. A single species, *H. gregarius*, has been found in the badlands of the western Great Plains. This species was a rather successful carnivoran, and by the early Oligocene (about 34 Ma), it had given rise to a few new species. Some of these species went on to establish more advanced groups of hesperocyonines and other canids. One of them, *Archaeocyon pavidus*, in turn gave rise to the great subfamily Borophaginae, and another is a primitive species of *Leptocyon* (we tentatively call it *Leptocyon* sp. A; its fossil record is too poor to permit us to give it a formal scientific name). Species of *Leptocyon* are the earliest members of the subfamily Caninae, which includes all the living canids. Therefore, *Hesperocyon* has played a central role in the history of the Canidae and is ancestral to all the subsequent canids.

Through a transitional species, *Hesperocyon coloradoensis*, the *Mesocyon–Sunkahetanka–Enhydrocyon* group arose and began to diversify. This lineage of the hesperocyonines was the most successful, including 11 species spanning the Oligocene plus the beginning of the Miocene (34 to 24 Ma). The evolutionary trend of this lineage was toward ever larger body size and more hypercarnivorous habits. The premolars became increasingly more robust, to the point of being able to crush bones. The carnassial teeth, especially the lower first molar, became more sharp-bladed, with long cutting edges and a trenchant talonid (figure 3.2), and the molars behind the carnassial teeth were reduced.

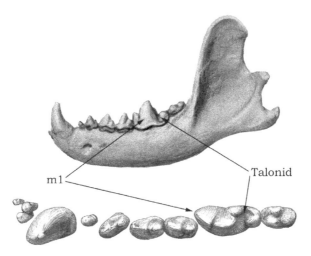

FIGURE 3.2
Sunkahetanka

Lateral and occlusal views of the left mandible of the hesperocyonine dog *Sunkahetanka*, showing the lower first molar (m1), or lower carnassial, with its distinctive talonid. Total length of skull: 15 cm.

At the base of this hypercarnivorous lineage are three species of *Mesocyon*, coyote-size predators that survived for more than 10 million years. Next to arise was *Cynodesmus*, which was slightly larger and stronger and trended toward a hypercarnivorous adaptation. By the late Oligocene (29 to 24 Ma), two genera, *Sunkahetanka* and *Philotrox*, represented the next stage of evolution in the *Mesocyon–Enhydrocyon* group. In size and morphology, these two genera are intermediate between *Cynodesmus* and *Enhydrocyon*. Finally, the genus *Enhydrocyon* emerged in the late Oligocene to represent the first fully hypercarnivorous canid capable of cracking large bones to feed on the marrow. With a moderate diversity of four species living in much of western North America, *Enhydrocyon* became extinct in the early Miocene (22 Ma), ending the line of a very successful group of hesperocyonines that dominated early carnivoran communities.

The genus *Osbornodon* split off from the *Mesocyon–Sunkahetanka–Enhydrocyon* group in the early Oligocene; it is the only group of hesperocyonines to show some hypocarnivorous tendencies. Early *Osbornodon* was represented by two small species in the early Oligocene. After a long gap in the fossil record in the middle to late Oligocene, advanced species of *Osbornodon* began to appear in the early Miocene and ultimately became extinct by the middle Miocene (around 15 Ma).

Beside these dominant groups of hesperocyonines, three smaller groups shared the adaptive range of early diversification within canids. *Paraenhydrocyon* is a small lineage of three species that has been found in the rocks of the northern Great Plains from the middle Oligocene through the early Miocene (32 to 22 Ma). This genus of medium-size canids independently evolved hypercarnivorous dental features (trenchant carnassials) that parallel those in *Enhydrocyon*. However, *Paraenhydrocyon* never acquired the massive, bone-cracking premolars found in *Enhydrocyon*. The peculiar genus *Caedocyon* suddenly appeared in the late Oligocene of Wyoming, perhaps an early offshoot of *Paraenhydrocyon*. A single specimen of this extremely rare species (*C. tedfordi*) is known in the fossil record, and much remains to be learned about this enigmatic form.

Finally, another peculiar hesperocyonine, *Ectopocynus*, occurs in the fossil record in the middle Oligocene (32 Ma) of the Great Plains. Members of this group have a unique array of premolars that feature massive, rounded, single-cusped premolars, in contrast to other hesperocyonines, whose premolars often have multiple cusps. *Ectopocynus* has three species, and fossils of this rare lineage can be found only in the large collection of fossil canids in the American Museum of Natural History in New York. The earliest and most primitive species, *E. antiquus*, is only slightly larger than *Hesperocyon*. Following *E. antiquus* is a transitional species, *E. intermedius*, which ultimately led to the terminal species, *E. simplicidens*. The last terminates the *Ectopocynus* line in the fossil record, from the early Miocene (about 18 Ma).

The following selected genera of hesperocyonines serve to indicate some representative lineages. They are generally organized from earlier and more primitive forms to later and more advanced forms (appendix 2).

PROHESPEROCYON

Estimated to have weighed less than 1 kg, *Prohesperocyon* (from Latin *pro* [before] and *Hesperocyon*), an early precursor of the canid family, possesses some of the primitive features seen in the family Miacidae, an archaic group of ancestral carnivores that probably gave rise to the suborder Caniformia (including bears [Ursidae], raccoons [Procyonidae], and weasels [Mustelidae]). Its bony middle-ear cover (tympanic bulla), which is the hallmark of living families of carnivorans, indicates its membership in the Canidae. Known by a single skull and lower jaw, this primitive dog was found in late Eocene (around 36 Ma) rocks of the southern Big Bend National Park in Texas.

HESPEROCYON

Weighing around 1 to 2 kg, *Hesperocyon* (from Greek *hesper* [western] and *cyon* [dog]), a small fox–size carnivore, is ancestral to all later canids. This enduring lineage of canid species lived for more than 10 million years, from the late Eocene to the late Oligocene (about 40 to 29 Ma) (figure 3.3). *Hesperocyon* has the unmistakable canid characteristics of a bony middle-ear covering that is partially divided by a thin, bony wall (septum). A hardened chamber, as contrasted to chambers formed by the less rigid cartilage in earlier carnivores, apparently can help animals to hear more acutely, an advantage that almost all living carnivores later acquired. *Hesperocyon* was the most abundant small predator of its time, and it is not uncommon to find good fossil records of this small dog in museum collections. This ubiquitous presence is also the reason Edward Drinker Cope, a preeminent vertebrate paleontologist in the late nineteenth century, named one of the most abundant species of this genus *Hesperocyon gregarius*, alluding to its possible social aggregation in life. Its hands and feet still had five fingers and toes, in contrast to the loss of the first finger (thumb) and big toe in more advanced canids (figure 3.4). The finger and toe bones became more compact, signaling a more upright posture, although *Hesperocyon* was probably not yet fully digitigrade (chapter 4). With a relatively deep claw, this small predator was apparently still capable of climbing trees. Fossil coprolites (fossilized remains of feces) presumed to belong to *Hesperocyon* contain bones of small rodents and rabbits (such as *Palaeolagus*), indicating that its diet included small vertebrates.

MESOCYON

Weighing between 6 and 7 kg, species of *Mesocyon* (from Latin *meso* [middle] and Greek *cyon* [dog]) were small, coyote-size carnivorans (figure 3.5), and they represent the first move among the canids to achieve a modest size and to specialize in meat eating (hypercarnivory [chapter 4]). Through transitional forms such as *Cynodesmus* and *Sunkahetanka*, *Mesocyon* gave rise to a group of fierce predators, *Enhydrocyon* species. The *Mesocyon* lineage had a long life span, from the early Oligocene to the early Miocene (34 to about 21 Ma). The best-known fossils of *Mesocyon* are from the John Day Fossil Beds National Monument in central Oregon, although they are also known in southern California and the northern Great Plains.

ENHYDROCYON

Weighing around 10 kg or more, *Enhydrocyon* (from *Enhydra*, a sea otter genus, referring to its otterlike skull shape, and Greek *cyon* [dog]) was the first canid to achieve a relatively large size and heavily built skull (figures 3.6 and 3.7). With its massive premolars, it probably was capable of crushing the bones of its prey to reach the nutritious marrow inside the bone cavity. As the first canid to become a

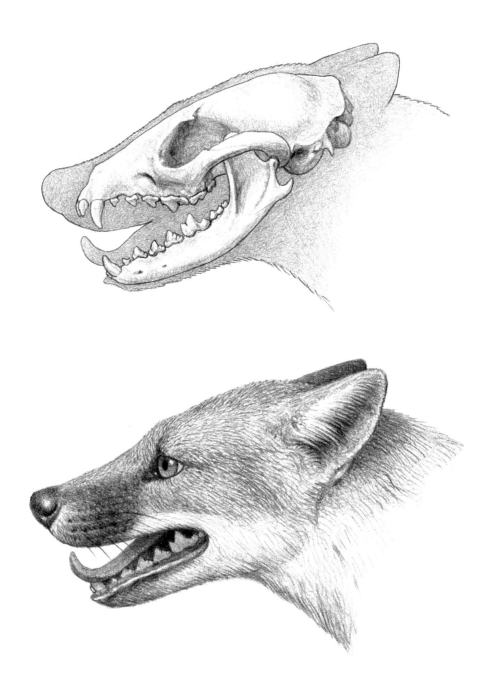

FIGURE 3.3

Hesperocyon gregarius

Skull, mandible, and reconstructed head of *Hesperocyon gregarius*. Because of its small size, pointed muzzle, and large orbits, the skull of *Hesperocyon* resembles in general shape the skull of the ancestral miacids, but also that of a small fox, a civet, or a genet. Reconstruction based largely on specimen USNM 437888 from Wyoming (32 Ma) and complemented from other specimens. Approximate total length of skull: 8 cm.

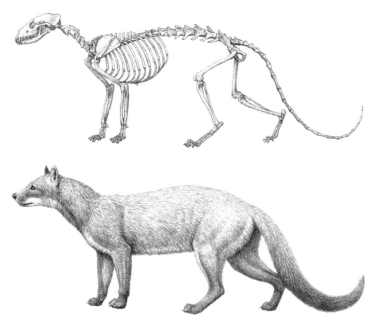

FIGURE 3.4

Hesperocyon gregarius

Skeleton and reconstructed life appearance of *Hesperocyon gregarius*, based mostly on specimen AMNH 8774 from Colorado (32 Ma). The standing posture is based on that of modern carnivores with similar skeletal morphology, especially civets and genets (family Viverridae). Like these carnivores, *Hesperocyon* differed from modern canids in having shorter forelimbs, especially the distal (lower) segments, with the metacarpals (the bones of the palm of the hand) being quite short. Civets and genets are plantigrade when walking on tree branches, but become digitigrade when walking on the ground, although even then their hind feet touch the ground at an angle rather than vertically, as they do in modern dogs. This posture is sometimes termed semiplantigrade. Although *Hesperocyon* would closely resemble a genet in body posture and proportions, the hypothetical coat pattern attributed here is based on patterns widespread among extant members of the Caniformia, the suborder to which the dog family belongs. Spotted patterns like that of genets are totally absent among dog relatives, and it seems safe to infer some sort of plain pattern for *Hesperocyon* as well. Reconstructed shoulder height: 20 cm.

FIGURE 3.5

Mesocyon coryphaeus

Reconstructed life appearance of *Mesocyon coryphaeus*. The head reconstruction is based largely on the holotype skull AMNH 6859 from Oregon (28 Ma). For lack of adequate associated postcranial fossils, the body proportions are based on fossils of the closely related species *Sunkahetanka geringensis*. Reconstructed shoulder height: 40 cm.

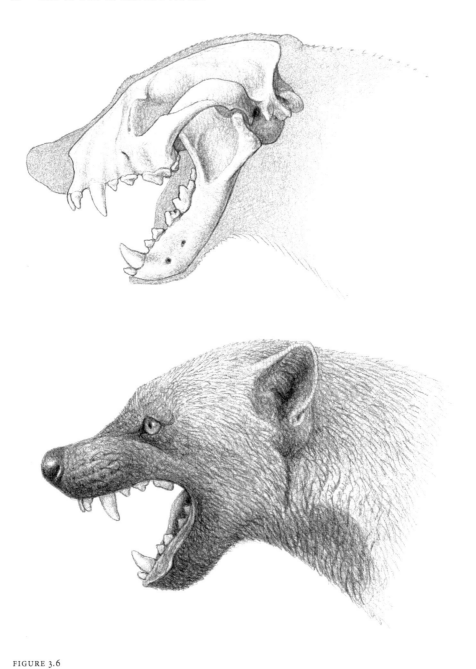

FIGURE 3.6

Enhydrocyon crassidens

Skull, mandible, and reconstructed head of *Enhydrocyon crassidens*, based on the holotype skull AMNH 12886 from South Dakota (28 Ma). Approximate total length of skull: 19 cm.

dominant carnivore, *Enhydrocyon* thrived from the middle Oligocene to the early Miocene (29 to 21 Ma).

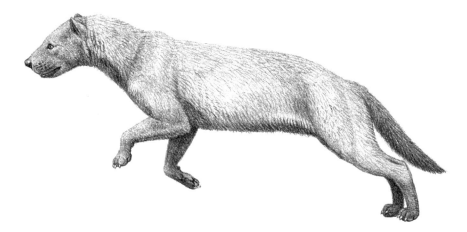

FIGURE 3.7
Enhydrocyon crassidens

Reconstructed life appearance of *Enhydrocyon crassidens*, based mostly on the holotype partial skeleton AMNH 12886. Although *E. crassidens* was considerably shorter at the shoulders than a modern jackal, it was a heavier animal, with a massive head and relatively short legs and neck. Reconstructed shoulder height: 44 cm.

OSBORNODON

The *Osbornodon* (from the name "Osborn" and Greek *don* [tooth]; named in honor of Henry Fairfield Osborn, founder of the Vertebrate Paleontology Department at the American Museum of Natural History in New York) lineage varies in size from that of a small fox to that of a large wolf. The first two species of this lineage, *O. renjiei* and *O. sesnoni*, occurred in the early Oligocene (about 33 to 29 Ma). They had a more omnivorous diet (hypocarnivory [chapter 4]), and their lower molars were slightly enlarged to handle a greater variety of food. Such an adaptation is more commonly seen in later canids, such as the borophagines and canines. After a long gap in the fossil record of nearly 8 million years, advanced species of *Osbornodon* appeared in the early Miocene (around 21 Ma) of the western United States. Weighing almost 20 kg, *O. fricki* was the largest species among the hesperocyonines (figures 3.8 and 3.9). The genus's body size had reached a critical point that allowed it to attack prey larger than itself. As often is the case with large predators (hypercarnivores [chapter 4]), this species was also the last member of the subfamily Hesperocyoninae in North America. *Osbornodon* last lived 15 Ma and has been found in middle Miocene rocks of the western United States.

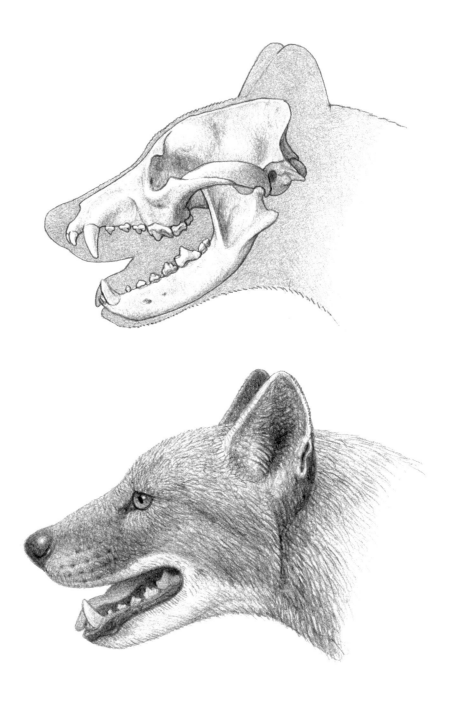

FIGURE 3.8

Osbornodon fricki

Skull, mandible, and reconstructed head of *Osbornodon fricki*, based on the holotype skull AMNH 27363 from New Mexico (15 Ma) and the skull F:AM 67098 from Nebraska (15 Ma). Approximate total length of skull: 23 cm.

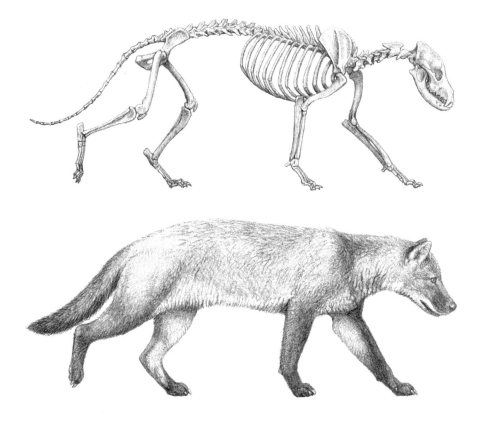

FIGURE 3.9

Osbornodon fricki

Skeleton and reconstructed life appearance of *Osbornodon fricki*, based mostly on the holotype skeleton AMNH 27363 from New Mexico (15 Ma). Proportions of the neck are based on the closely related species *O. brachypus*. The length of the tibia is unknown, but is inferred to be short in line with the short forearm. Compared with a modern wolf of similar body mass (as inferred from the thickness and articular areas of the long bones), the limbs of *O. fricki* had shorter distal (lower) elements, resulting in a moderate shoulder height of about 55 cm. The neck was a little shorter, and some of its muscle insertions were more pronounced. In life, *O. fricki* would have looked somewhat like a short-legged wolf.

Borophagines

The subfamily Borophaginae is named after the genus *Borophagus*, a terminal group of large canids that developed strong bone-crushing dentitions. Because of this highly specialized dental adaptation for cracking bones, the borophagines have been known as the "hyaenoid dogs." However, many borophagines were not bone crushers. The borophagines first appeared about 32 Ma and declined to extinction about 2 Ma, spanning the middle Oligocene to the Pliocene. During the 30 million years of their existence, they were confined to the North American continent. In the history of the canids, the borophagines are the most diverse, occupying the widest ecological niches—ranging from the hypercarnivorous, supersize *Epicyon* and *Borophagus* to such hypocarnivorous, small- to medium-size omnivores as *Cynarctoides*, *Phlaocyon*, and *Cynarctus*. At the peak of their diversity in the middle Miocene (about 16 to 12 Ma), as many as 15 species of borophagines occupied the entire North American continent (as compared with six or seven species of living canids in North America), becoming the dominant group of carnivores for several million years.

The borophagines had their beginning in the form of a small, fox-size genus named *Archaeocyon*, which arose from *Hesperocyon*. *Archaeocyon* acquired a basined talonid in the lower first molar, formed by two small cusps (hypoconid and entoconid), as well as corresponding modifications on the upper molars, such as the presence of an extra cusp called a *metaconule*. These dental modifications were inherited by all subsequent lineages in the borophagine subfamily and are useful features for recognition of this group (figure 3.10). Roughly contemporaneous with *Archaeocyon* were a few fox-size canids such as *Oxetocyon*, *Otarocyon*, and *Rhizocyon*. These basal borophagines permit a glimpse of the early diversification of the borophagines. They tended to be small, hypocarnivorous predators with a generalized diet of invertebrates, fruits, and small vertebrates.

After the initial diversification, early borophagines remained small to medium size during much of the first half of their history in the Oligocene through the early Miocene (34 to 17 Ma). Many, such as *Cynarctoides* and *Phlaocyon*, developed dental specializations that indicate an omnivorous diet, rather like modern raccoons. These two genera form the tribe Phlaocyonini as the first hypocarnivorous group in canid history. The next step of borophagine evolution saw two small transitional forms, *Cormocyon* and *Desmocyon*, both with good fossil records throughout the western United States. Following these transitional genera was yet another group of omnivores represented by *Cynarctus* and *Paracynarctus*, which are included in the

Talonid of m1

FIGURE 3.10

Archaeocyon

Occlusal view of the lower-left dentition of the early borophagine *Archaeocyon*, showing the basined talonid on the lower carnassial (m1).

subtribe Cynarctina. The borophagines' early evolution thus appears to include a number of hypocarnivorous lineages that developed less predacious habits, as compared with their contemporaries, the hesperocyonines. The tendency to become less predacious may suggest competitive pressure from the larger, more dominant hesperocyonines, which had established themselves as top predators. The early borophagines may have been forced toward omnivore niches that were less competitive with those of the hesperocyonines.

The next phase of borophagine evolution saw a number of transitional genera moving toward larger and more hypercarnivorous forms mostly in the early to middle Miocene (19 to 15 Ma). Representatives of these canids include *Metatomarctus*, *Microtomarctus*, *Protomarctus*, and *Tephrocyon*. This series of intermediate-size canids generally evolved toward a more predacious lifestyle, and two genera, *Euoplocyon* and *Psalidocyon*, even acquired specializations toward a pure meat diet. Borophagines in this transitional phase were still too small to challenge large and established top predators, such as the advanced hesperocyonines and contemporaneous felids. To make this challenge, the borophagines had to wait until the next stage of their evolution—the subtribe Aelurodontina.

Starting with *Tomarctus*, borophagines quickly transitioned into powerful, hypercarnivorous predators, the Aelurodontina, in the middle to late Miocene (16 to 9 Ma). This first lineage of top predators within the borophagines was formed by two genera, *Tomarctus* and *Aelurodon*. The much restricted *Tomarctus* (as opposed to earlier genera that included a much wider array of species) went through only two intermediate species, *T. hippophaga* and *T. brevirostris*, before giving rise to the terminal genus, *Aelurodon*. With the rise of *Aelurodon*, which includes six species, the borophagines truly established themselves as a formidable predator rivaled by few others. Such a momentous event apparently followed the decline of the last hesperocyonine, *Osbornodon*.

The middle Miocene also saw the appearance of the final group of the borophagines, the subtribe Borophagina. This subtribe split from the Aelurodontina early in the middle Miocene and continued on until the late Pliocene (16 to 2 Ma). The early progenitors of this lineage belong to the genus *Paratomarctus*, a coyote-size, generalized predator. Another early genus, *Carpocyon*, split from *Paratomarctus* at the very beginning of the middle Miocene (about 16 Ma) and achieved a modest diversity of four species, one of which (*C. webbi*) could be as large as a living wolf. *Carpocyon* also sported somewhat hypocarnivorous dentition, suggesting a minor reversal toward a more omnivorous diet along a line of descent that was largely hypercarnivorous in direction.

At the beginning of the middle Miocene, the final borophagine group, the *Protepicyon*–*Epicyon*–*Borophagus* lineage, arose. As top predators of their time, members of this lineage reached the pinnacle of borophagine evolution. In particular, *Epicyon* went on to become the largest canid ever evolved and acquired a powerful set of bone-crushing teeth. Our record indicates that this group of hypercarnivores first evolved in California and New Mexico (*Protepicyon raki*) in the late Miocene. It soon spread to much of North America before it became extinct at the very end of the Miocene (around 5 Ma).

Finally, the great *Borophagus* lineage arose from an early species of *Epicyon*. *Borophagus* features some of the most extreme adaptations of bone-crushing dentitions. There are eight species in this great lineage of bone crushers (some were previously separated under the generic name *Osteoborus*). As did *Epicyon*, the first *Borophagus* (*B. littoralis*) appeared initially in the late Miocene (around 5 Ma) of California and quickly spread eastward to the rest of North America. With the exception of a diminutive species from Florida (*B. orc*), most species of *Borophagus* became progressively larger in size and more powerful in their bone-crushing teeth, culminating in the terminal species, *B. diversidens*. With the extinction of *B. diversidens* at the end of the Pliocene (around 2 Ma), the history of the subfamily Borophaginae ended, although there are unconfirmed suggestions that *Borophagus* may have survived into the Pleistocene in Mexico.

Descriptions and illustrations of selected genera of the borophagines offer greater knowledge of some representative lineages. They are organized from earlier and more primitive forms to later and more advanced forms (appendix 2).

ARCHAEOCYON

Very similar in size (less than 2 kg) and shape to *Hesperocyon*, species of *Archaeocyon* (from Greek *archae* [ancient] and *cyon* [dog]) are the ancestors of all other borophagines. Fossils of *Archaeocyon* also compose some of the earliest records of the subfamily Borophaginae. This genus is represented by two species that have a relatively long fossil record. Spanning the middle to late Oligocene (32 to 24 Ma), *Archaeocyon* fossils can be found in the northern Great Plains and the western coast of North America (figures 3.11 and 3.12).

OTAROCYON

Otarocyon (from Greek *otaros* [large-eared] and *cyon* [dog]) is one of the smallest canids, but it has an outsized bulla, the bony chamber that houses the middle ear bones (figure 3.13). A similarly enlarged bulla can be found in the modern fennec fox (*Vulpes zerda*), of the Sahara and Arabian deserts, which also sports a large, erect external ear. Such a large middle-ear chamber may be an adaptation for

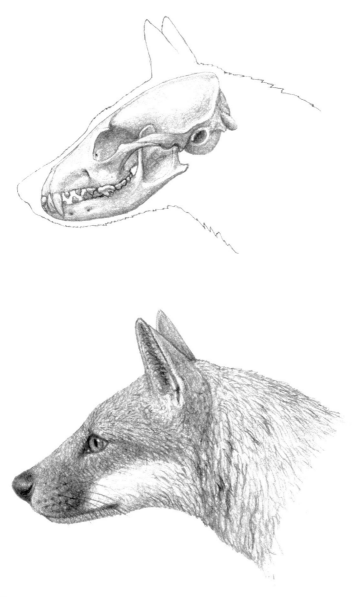

FIGURE 3.11
Archaeocyon leptodus

Skull, mandible, and reconstructed head of *Archaeocyon leptodus*, based on several specimens, chiefly F:AM 63971 from Nebraska (27 Ma). Approximate total length of skull: 10 cm.

enhanced hearing of low-frequency sounds in desert environments, and it is quite possible that *Otarocyon*'s inflated middle-ear region may also be associated with a large external ear as an adaptation for open environments. *Otarocyon* also has the distinction of being the oldest borophagine lineage, living as early as the early

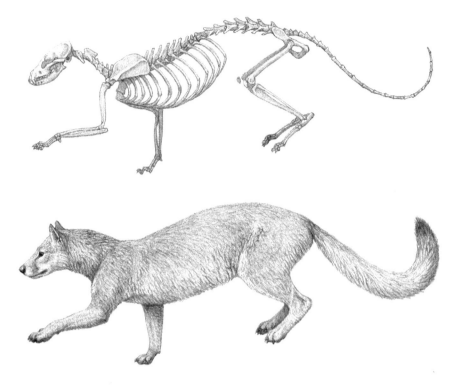

FIGURE 3.12

Archaeocyon leptodus

Skeleton and reconstructed life appearance of *Archaeocyon leptodus*, based mostly on skeleton F:AM 49060 from Wyoming (27 Ma). In size, body proportions, and posture, *Archaeocyon* closely resembled *Hesperocyon*, as is evident in this reconstruction. These primitive dogs resembled modern genets in having relatively long tibiae and hind feet, suggesting a well-developed jumping ability, so, overall, they must have been extremely agile, versatile little creatures. Reconstructed shoulder height: 27 cm.

Oligocene (34 Ma). Weighing less than 1 kg, this small, fox-size borophagine lived only in what are now the northern states of South Dakota, Wyoming, and Montana.

CYNARCTOIDES

The genus *Cynarctoides* (from *Cynarctus*, a borophagine genus, and the Greek suffix *-oides* [similar]) features a peculiar set of teeth that mimics the conditions in the ringtail (*Bassariscus astutus*). *Cynarctoides* also developed crestlike structures in its teeth that are called *selenodont* in artiodactyls (animals such as sheep and cows). Such a structure is designed for processing plant food by herbivores, and it rarely occurs in carnivorans. In fact, *Cynarctoides* is the only caniform (doglike) carnivoran that possesses such a peculiar structure. It thus raises the tantalizing question whether *Cynarctoides* was an herbivore. Dedicated herbivory in a carnivoran is

FIGURE 3.13

Otarocyon cooki

Skull, mandible, and reconstructed head of *Otarocyon cooki*, based on specimen F:AM 49020 from Wyoming (27 Ma). The huge auditory bullae of this small dog suggest the presence of very large external ears. The resemblance of the reconstructed animal to a living fennec fox is strengthened by the general proportions of the head, with its large eyes and very small, pointed muzzle. Such features would give this canid an unmistakable, expressive face. Approximate total length of skull: 6 cm.

exceedingly rare and seen only in the living giant pandas of southern China. Seven species of *Cynarctoides* flourished in much of western North America, and all of them remained small in body size, no more than 1 kg in weight.

PHLAOCYON

Species of *Phlaocyon* (from Greek *phlao* [eat greedily] and *cyon* [dog]) are another lineage of hypocarnivorous canids. Living from the middle Oligocene to the middle Miocene (30 to 16 Ma), the nine species in the genus *Phlaocyon* ranged from the size of a kit fox to that of a large coyote (figure 3.14). The small- to medium-size members (less than 5 kg) were rather raccoonlike in their bunodont dentition (similar to human dentition) and were likewise omnivorous in their diet (figure 3.15). Two large species (around 15 kg), however, reversed this trend to become hypercarnivorous with robust jaws. Earlier paleontologists had mistakenly placed *Phlaocyon* in the raccoon family, Procyonidae, because of its raccoonlike dental anatomy. It is now widely recognized that the raccoonlike dental features were the result of independent evolution. Independent achievement of similar dental anatomy is quite com-

FIGURE 3.14

Phlaocyon leucosteus

Reconstructed life appearance of *Phlaocyon leucosteus*, based mostly on the holotype skeleton AMNH 6768 from Colorado (28 Ma). This small dog was comparable in body size to *Archaeocyon*, but it had somewhat different body proportions, with shorter distal (lower) segments in the limbs. The tibia, or shinbone, in particular was markedly shortened in *Phlaocyon*. The choice of a facial mask for the reconstructed coat is based on the observation by zoologists Chris Newman, Christine D. Buesching, and Jerry O. Wolff (2005) that markings on the face were a feature independently evolved by a wide range of "midguild" carnivores (including raccoons, badgers, civets, and the "raccoon dog," among others) as a warning to deter attacks from larger predators. *Phlaocyon*, with its unimpressive size and its likely omnivorous diet, is a reasonable candidate to have evolved such markings. Reconstructed shoulder height: 22 cm.

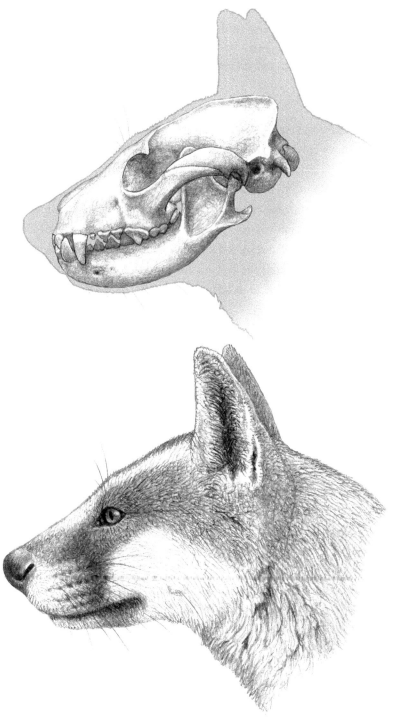

FIGURE 3.15

Phlaocyon leucosteus

Skull, mandible, and reconstructed head of *Phlaocyon leucosteus*, based on the holotype skull AMNH 6768 from Colorado (28 Ma). Approximate total length of skull: 9 cm.

mon in the evolution of carnivorans and provides great examples of the convergent development of similar features in different evolutionary lineages because of their similar functional requirements.

CYNARCTUS

Cynarctus (from Greek *kynos* [dog] and Latin *arcto* [bear]) belongs to a group with extremely hypocarnivorous teeth (figure 3.16). In advanced members, the molars were substantially enlarged and developed numerous small cusps quite similar to those in the bear family, Ursidae. *Cynarctus* and its sister lineage *Paracynarctus* lived mainly in the Miocene (19 to 9 Ma), on average weighed less than 15 kg, and achieved a modest diversity of five species. Besides *Cynarctus*'s usual occurrences in western North America, *Paracynarctus* was also one of the few genera of borophagines that managed to inhabit the Eastern Deciduous Forest along the east coast of North America, possibly due to its adaptations to more wooded environments than is usual for most canids.

TOMARCTUS

In the past, many medium-size borophagines were included in the genus *Tomarctus* (from Latin *tom* [cut] and *arcto* [bear]), so it did not constitute a natural lineage (a genealogically related group that includes its ancestors as well as *all* its descendents) as traditionally conceived. It was a taxonomic wastebasket into which many questionable species were tossed. Here we adopt a taxonomy that greatly reduces the members of *Tomarctus* to only two species leading to *Aelurodon*. These two species of *Tomarctus* lived in the middle Miocene (16 to 14 Ma) of western North America and weighed approximately 14 to 18 kg. Many other species formerly included in *Tomarctus* are now given distinct generic identifications of their own. For a long time, *Tomarctus* was thought to be related to the living *Canis* due to its possession of a basined talonid in the lower first molars. This idea was perpetuated in the popular literature. Our own studies, however, have traced the basined talonid back to the origin of both borophagines and canines. Therefore, this feature cannot be a manifestation of a special relationship between *Tomarctus* and *Canis*. In fact, we now know that the relationship between the two genera is very distant.

AELURODON

This group of hypercarnivores is the first among borophagines to develop jaws strong enough and teeth heavy enough to be able to crack bones. Weighing between 20 and

FIGURE 3.16
Cynarctus

Skull, mandible, and reconstructed head of *Cynarctus*, based on specimen F:AM 49172 from Nebraska (10 Ma). Approximate total length of skull: 20 cm.

40 kg, *Aelurodon* (from Greek *ailurus* [cat] and *don* [tooth]) arose from *Tomarctus* around 16 Ma in the form of *A. asthenostylus* and ended in a giant form, *A. taxoides*, about 9 Ma (figure 3.17). There are many morphological similarities between *Aelurodon* and the living African hunting dog (*Lycaon pictus*); for example, they share broad palates, multicuspid premolars, and trenchant cutting teeth (carnassials) (figure 3.18). Because living hunting dogs are social hunters (chapter 5), it is tempting to speculate that *Aelurodon* may have been capable of social hunting as well.

FIGURE 3.17

Aelurodon ferox

Reconstructed skeleton, musculature, and life appearance of *Aelurodon ferox*, based mostly on partial skeleton F:AM 61746 from Nebraska (10 Ma), completed with material from two partial skeletons from Nebraska and New Mexico. Apart from the shorter neck and slightly shorter distal (lower) limb segments, *A. ferox* was similar to a modern gray wolf in size and body proportions. Unlike modern dogs, *Aelurodon* and all other borophagines had a small but recognizable first digit (indicated by the arrow in the skeletal drawing) on their hind feet. Reconstructed shoulder height: 75 cm.

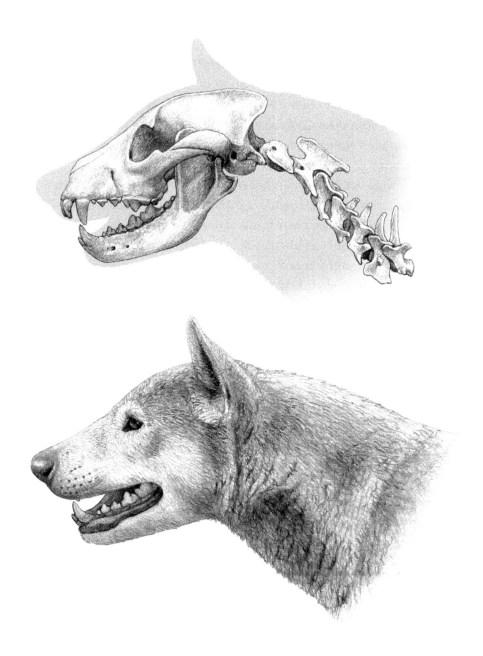

FIGURE 3.18

Aelurodon mcgrewi

Skull, cervical vertebrae, and reconstructed life appearance of the head of *Aelurodon mcgrewi*, based on the holotype partial skeleton F:AM 22410 and on skull F:AM 61778, both from Nebraska (11 Ma). With its robust skull, high maxilla, and prominent sagittal crest for the insertion of powerful temporalis muscle, the head of *Aelurodon* would have resembled the heads of the most predaceous among modern dogs, such as the Indian dhole and the African hunting dog, but would have been even more massive, bringing to mind modern hyenas as well.

Most members of *Epicyon* (from Greek *epi* [near, before] and *cyon* [dog]) ranged between 30 and 75 kg. The largest *Epicyon* rivaled a brown bear in size. This made them truly the top dogs of their day. In fact, no other canid has exceeded this body size. Arising from a lineage called *Protepicyon*, *Epicyon*, like some hyenas, had an enlarged lower fourth premolar, enabling it to deliver a powerful bite (figure 3.19). Concentrating bite force on this tooth would crack bones open to extract the nutritious bone marrow inside. This lineage (including its progenitor, *Protepicyon*) had a long history, living 16 to 7 Ma in the Miocene, and was found across western North America. During much of this period, a pair of *Epicyon* species coexisted side by side in many of the localities where fossils of this genus have been found: *E. saevus* and *E. haydeni*, with the latter always being the larger of the two (figures 3.20 and 3.21). This pair's competitive coexistence resulted in a situation marked by character displacement, a biological phenomenon that permits species to coexist through a substantial size difference between them so that they can avoid competing for the same resources. In the present case, *E. saevus* kept its distance from *E. haydeni* by being smaller and thus seeking somewhat different food resources (prey of smaller size). Interestingly, although both species evolved toward larger body sizes, they continued to maintain their "distance" from each other by keeping a relatively constant body-size difference.

The genus *Borophagus* (from Greek *boros* [voracious] and Latin *phago* [eating]) is the name bearer of the entire subfamily Borophaginae, and the notion of a hyena-like dog is closely associated with this group (figure 3.22). Weighing 20 to 40 kg or more, *Borophagus* was the last member of the subfamily. It first appeared along the Pacific coast in the later middle Miocene (about 12 Ma). Derived from *Epicyon*, *Borophagus* carried to the extreme the enlarged fourth lower premolars in *Epicyon* and was another efficient bone cracker. To further assist in the cracking of bones, it had a strong jaw and a domed forehead that helped to strengthen the jaw mechanism in order to sustain the tremendous force required for crushing bones (figure 3.23). At the beginning of the late Miocene (about 9 Ma), it had spread to the rest of North America, including Florida and central Mexico. *Borophagus* became extinct just before the beginning of the Ice Age (Pleistocene [1.8 Ma]), thus ending the 30-million-year existence of the great borophagine lineage.

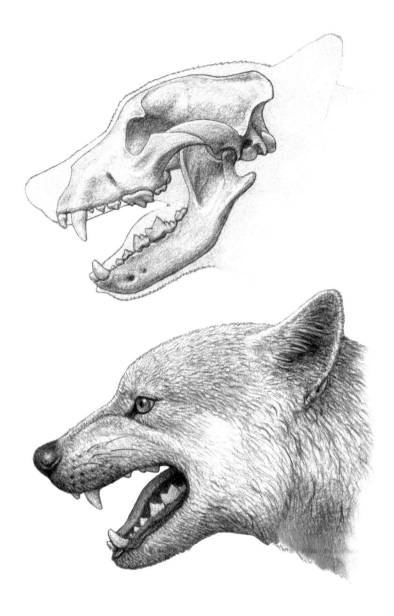

FIGURE 3.19

Epicyon haydeni

Skull, mandible, and reconstructed head of *Epicyon haydeni*, based on partial skull F:AM 61474 and other specimens, all from Kansas (10 Ma). Approximate total length of skull: 34 cm.

FIGURE 3.20

Epicyon saevus

Reconstructed life appearance of *Epicyon saevus*, based on partial skeleton AMNH 8305 from Nebraska (10 Ma), among other specimens. Reconstructed shoulder height: 56 cm.

FIGURE 3.21

Epicyon haydeni

Reconstructed life appearance of *Epicyon haydeni*, based on several specimens from Kansas. Reconstructed shoulder height: 90 cm.

FIGURE 3.22

Borophagus diversidens

Reconstructed life appearance of *Borophagus diversidens*, based on skeleton MSU 8034 from Texas (4 Ma). Although *Borophagus* is often known as a hyena-like dog because of its extreme cranial adaptations for bone cracking, the remarkably complete skeleton from Blanco Canyon, Texas, clearly shows that it did not have the short hind limbs and sloping back of modern hyaenids, but instead had body proportions more typical of dogs, with long hind limbs and a level back. Reconstructed shoulder height: 62 cm.

Canines

Members of the subfamily Caninae are the only representatives of the family Canidae still living, although the family's history begins with the Borophaginae. Both groups arose in the early Oligocene (32 Ma) of North America (appendix 2). Like the borophagines, the canines were confined to North America until the very end of the Miocene (6 Ma). At that time, geological events united Asia and North America at the Beringian land bridge (now the Bering Strait); then uplift of middle America formed the Isthmus of Panama 3 million years later, uniting the Americas. Canids took advantage of both bridges to disperse to new continents. These events allowed access to new prey in large temperate and tropical regions of the world, and the canids reacted to these advantages with spurts of evolution in the new environments.

In the late Miocene, canines were just beginning a major diversification (adaptive radiation) after having spent most of the preceding 23 million years as small, generalized members of their initial stock (*Leptocyon* species). They coexisted with larger hesperocyonines and the more vigorously evolving borophagines during most of

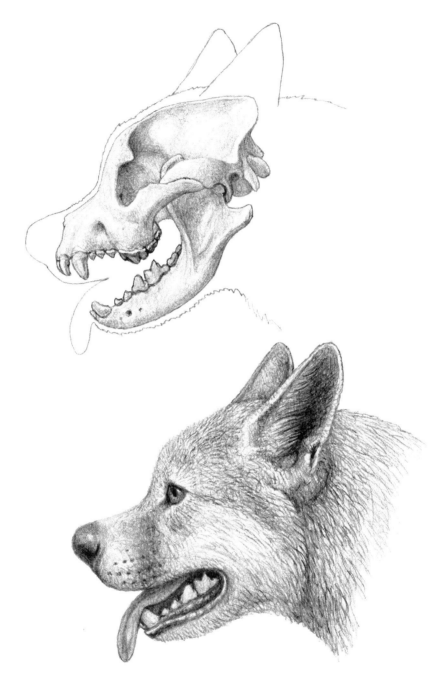

FIGURE 3.23

Borophagus secundus

Skull, mandible, and reconstructed head of *Borophagus secundus*, based on specimens F:AM 23350 from Texas (6 Ma) and AMNH 18919 from Nebraska (6 Ma). The *Borophagus* skull shows the most pronounced adaptations for bone cracking among the borophagines, including a shortened muzzle and a strongly domed forehead. Approximate total length of skull: 21 cm.

the Miocene, a coexistence that oriented canines' adaptive possibilities for most of their existence.

Species of *Leptocyon* were fox-size canids with a varied diet of small animals and fruit. Eleven species have been recognized so far. The great similarity in their dentition and overall size indicates that their basic adaptation to a mixed diet changed little during the Miocene. They closely resembled early borophagines, such as *Archaeocyon*, except for their longer and shallower jaws and their delicate teeth, which were more useful in capturing small, active prey than in administering a killing bite, as did their borophagine competitors. Species of the genus *Vulpes*, descendants of *Leptocyon*, continued this adaptation, but some species, such as the arctic fox (*V. lagopus*), paralleled the borophagines in having a larger dentition set in a deeper jaw in order to deal with arctic hares only a little smaller than the fox.

Around 10 Ma, species allied in the genus *Vulpes* were joined by species of the fossil genus *Metalopex*, allied to the living gray fox, *Urocyon*. In both *Metalopex* and *Urocyon*, the crushing elements of the carnassials and molars were enlarged in adaptation to feeding on vegetation and insects, again paralleling the Miocene *Cynarctoides* species among the borophagines. The addition of hypocarnivory among the canines was the first sign of enlargement of the range of adaptation within the Caninae, and from the late Miocene to present times other ways of life appeared and evolved. A widespread new adaptation occurred at this time with the extension of the nasal cavity into the frontal bone. This expansion of the nasal cavity behind the turbinates—the organs involved in heat and moisture exchange and in the sense of smell (chapter 4)—was apparently related to forces acting on the skull from the enlargement of the dentition. However, even fox-size canines possessed such sinuses, so the adaptation was not altogether size related. It was, however, confined mostly to the more carnivorous and hypercarnivorous forms, which suggests that it may have been related to stresses placed on the skull during capture of and feeding on prey similar in size or larger than themselves. This adaptation may have been key in finally allowing the Canidae to deal with prey of increasing size.

The earliest Caninae to reach eastern Asia across Beringia did so during a globally warm span of time in the early Pliocene (5 to 4 Ma), when sea level would have been relatively high (although canines apparently arrived in Europe and Africa as early as the late Miocene [6 Ma] [chapter 7]). This warming event resulted in submerged continental shelves, but the emergence of the Beringian land bridge was also affected by zones of repeated uplift, as is the Bering Strait today. The canines that dispersed to Asia along this path are well recorded in fossil deposits in northern China. Included in these deposits are fossil species of the living genus of raccoon dog (*Nyctereutes*), now confined to eastern Asia. These animals appear to be closely related to the crab-eating fox, or zorro (*Cerdocyon*), now restricted to South America, but whose earliest fossil record is in the late Miocene of North America.

Accompanying the raccoon dog was *Eucyon*, a genus of canid closely related to *Canis* itself and now extinct. In fact, the earliest Asian species, *E. davisi*, was identical to its contemporary in North America. In the early Pliocene (5 Ma), *E. davisi* became extinct in North America, but survived in Asia, where it underwent further evolution and dispersal across the continent, persisting there nearly to the end of the Pliocene (2.6 Ma) and giving rise to new species. Small species of *Canis* also appeared in Asia in the early Pliocene. They were coyote- or jackal-size animals, although not closely allied to the jackals of today. Similar species were also present in North America. By the end of the Pliocene, larger, wolf-size species of *Canis* appeared in China and spread westward across the continent. In North America, however, native *Canis* species remained the size of coyotes until the last borophagine (*Borophagus diversidens*) became extinct. Eurasia continued to contribute products of its canid evolution to North America during the Pleistocene ice ages. Even the largest of all species of *Canis*, the dire wolf (*C. dirus*), arose from an immigrant, *C. armbrusteri*. Finally, the gray wolf (*C. lupus*) reached midcontinent North America in the last glacial cycle (100,000 years ago) after a much longer residence in North America above the Arctic Circle (since 500,000 years ago at least).

Because the present geographic relationship between North America and South America was established only 3 Ma and the path to the south lay across the tropics, a region with a limited fossil record, we have little direct evidence of the biotic history of Middle America. What we do have comes from deposits along the border of the United States and Mexico and then far to the south in temperate South America.

Morphological (Tedford, Taylor, and Wang 1995) and molecular (Wayne 1993; Wayne et al. 1997) studies of the evolutionary relationships among the living South American canids (subtribe Cerdocyonina) have shown that these forms belong to a single closely related evolutionary group (that is, they are *monophyletic*) composed of several lineages. Large species of *Canis* (*C. dirus*) also occur in the South American fossil record, but they represent dispersals into South America during the Pleistocene (1.8 to 0.01 Ma) and are not part of the monophyletic group of species that we have designated the subtribe Cerdocyonina of the tribe Canina. We regard the living crab-eating fox (*Cerdocyon*) and its fossil species as typical of the subtribe. The South American fossil record indicates that a lineage of large hypercarnivores—including the recently extinct Falkland Island fox (*Dusicyon*) and species of the extinct genera *Protocyon* and *Theriodictis* that occurred in the Pleistocene of Andean and Amazonian South America—also belong to the subtribe. The subtribe Cerdocyonina surprisingly has an earlier record in the late Pliocene (2 Ma) of Florida, where the remains of a large form like *Theriodictis* were found. The maned wolf (*Chryso-*

cyon) is also present in the Pliocene of North America (northern Mexico and adjacent Arizona). In molecular studies, the extant species *Chrysocyon brachyurus* is often depicted as most closely related to the living bush dog (*Speothos*) of Amazonia. Extinct species of the crab-eating zorro (*Cerdocyon*) are also present in Pliocene deposits of Baja California and Arizona. All these North American taxa were present prior to the establishment of the Isthmus of Panama, indicating that much of the fundamental differentiation among the Cerdocyonina had taken place in North or Middle America before the appearance of the group in South America.

At this point, we can shift from generalities to the specifics of elements in the evolution of the tribe Canina. During the history of this tribe, canines gained access to adjacent continents before the end of the Ice Age (15,000 years ago); thus they occupied the entire New World as well as Eurasia and Africa and were rapidly diversifying. Products of this diversity frequently returned to their continent of origin. Nowhere was this diversification more dramatically demonstrated than in the invasions of North America, whose native canid fauna included only the coyote (*Canis latrans*), a large extinct coyote relative (*C. edwardii*), and smaller foxes (*Vulpes macrotis*, *V. vetus*, and *Urocyon cinereoargenteus*). In the invasion, North America acquired the red fox (*V. vulpes*), the arctic fox (*V. lagopus*), and the gray wolf (*C. lupus*), all immigrants from Eurasia. From time to time, large hypercarnivores of Eurasian origin—such as *C. armbrusteri*, species of the extinct genus *Xenocyon*, and species of the extant Asian dhole (*Cuon*)—appeared in midlatitude North America south of the Pleistocene glaciers.

LEPTOCYON

The genus *Leptocyon* (from Greek *leptos* [slender] and *cyon* [dog]) includes 11 species, all small animals weighing less than 2 kg. They are the most primitive canines. They share features with the borophagines, which indicates that they are sister groups, and they appear in the geological record at about the same time, in the early Oligocene (34 Ma). Both subfamilies are characterized by lower carnassials (the first molar) with a bicuspid talonid, but *Leptocyon* species had longer, lower jaws containing simpler premolars separated by gaps (figure 3.24) rather than the large premolars without gaps, as in the early borophagines such as *Archaeocyon*. The *Leptocyon* skull and dentition were adapted to snatching small, quick-moving prey, whereas borophagines were capable of a more powerful killing bite. Fossils of *Leptocyon* species show limited skeletal adaptations, but there are suggestions of relatively primitive and more advanced foxlike forms and even of small, kit fox–like species that are the smallest known canids (*L. delicatus*). Toward the end of their existence around 9 Ma, one lineage began to resemble species of the extant fox, *Vulpes* (figure 3.25).

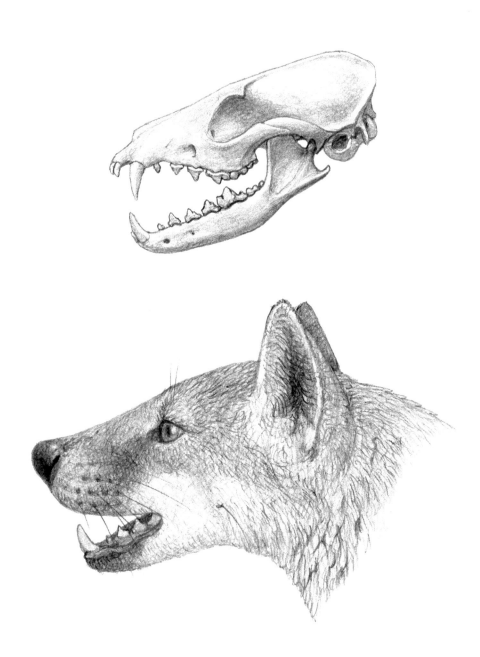

FIGURE 3.24

Leptocyon vafer

Skull and reconstructed head of *Leptocyon vafer* (9 Ma). Total length of skull: 11 cm.

FIGURE 3.25
Leptocyon vafer
Life reconstruction of *Leptocyon vafer*. Reconstructed shoulder height: 25 cm.

VULPES

The foxes, placed mostly in the genus *Vulpes* (Latin for "fox"), are diverse, and some are singled out in the nomenclature because of their unusual form (for example, the large-eared fennec fox [*Fennecus* or *Vulpes zerda* (figure 3.26)] and the arctic fox [*Alopex lagopus* or *Vulpes lagopus*]). They first occurred in North America in the late Miocene (about 9 Ma). Two species expanded their range into eastern Asia in the early Pliocene (4 Ma) to become the earliest members of a prodigious diversification among Eurasian and African canids, which returned to North America during the Pleistocene. Small species also modestly diversified in North America as well, but their remains are curiously rare in fossil collections.

CERDOCYONINA

The canid fauna of South America appears to represent closely related members of a phyletic group, the Cerdocyonina, the most diverse subtribe of canids to inhabit a single continent at the present time (figure 3.27). The fossil record of this group is just becoming known principally because its record in North America lies mostly in the southern latitudes from the Gulf Coast and southwestern United States into northern Mexico. The fossil record now available indicates that considerable differentiation among genera of this group took place in this region prior to the emergence of the Isthmus of Panama (about 3 Ma). Genera in North America before this event

FIGURE 3.26
Fennec fox
The extant fennec fox (*Vulpes zerda*). Shoulder height: 22 cm.

were species of the crab-eating zorro (*Cerdocyon*; from Greek *cerdos* [wily] and *cyon* [dog]), the long-legged maned wolf (*Chrysocyon*; from Greek *chrysos* [gold] and *cyon* [dog]) (figure 3.28), and a wolf-size hypercarnivore called *Theriodictis* (from Greek *therion* [beast] and *dictis* [eater]). These forms occur in various branches of the South American canid phyletic tree, suggesting that other closely related groups were also present, but their remains have not been discovered. South American fossil records after access to the continent was established include most of the known living species. The remains of additional extinct hypercarnivores are also present.

EUCYON

In the latest middle Miocene (10 Ma) of the western United States, a small canid about the size of a jackal (15 kg) appeared that is placed in its own genus, *Eucyon* (from Greek *eu* [primitive] and *cyon* [dog]), which phyletically stands between the South American canines and the many species of *Canis* to follow. Among other features characteristic of this group is the gradual increase in the size of the frontal sinus, which lies between the nasal region and the brain within the dorsal part of the skull. *Eucyon* has a frontal sinus like that of most South American canids, but not as

FIGURE 3.27

Bush dog

The extant bush dog (*Speothos venaticus*). Shoulder height: 25 cm.

FIGURE 3.28

Maned wolf

The extant maned wolf (*Chrysocyon brachyurus*). Shoulder height: 87 cm.

FIGURE 3.29
Eucyon davisi
Life reconstruction of *Eucyon davisi* (5 Ma). Reconstructed shoulder height: 38 cm.

expanded as is usually found in *Canis* species of similar size. This feature seems to be an adaptation for feeding as well as to the enlargement of the adult skull.

Eucyon has one geologically long-living species, *E. davisi* (figure 3.29), which existed during the late Miocene (7 Ma) in North America, when it joined a number of North American mammals invading eastern Asia during the early Pliocene (around 6–5 Ma), such as *E. zhoui* (figure 3.30). It persisted into the middle Pliocene (3 Ma) in Eurasia, a span of 3 million years. While in Eurasia, *Eucyon* underwent a modest radiation of Asian species that extended across that continent. In North America, *Eucyon* gave rise to *Canis* species at the close of the Miocene (6 Ma).

CANIS

Species that can be grouped in the genus *Canis* (Latin for "dog") first appeared in the late Miocene (6 Ma) in North America. Only a little larger than their ancestor *Eucyon davisi* and similar to it, their remains have been found in deposits exposed in the southwestern United States and adjacent Mexico. By the beginning of the Pliocene (5 Ma), a somewhat larger form, *C. lepophagus*, arose in the same region; by the early Pleistocene (1 Ma), the living coyote (*C. latrans*) was present (figure 3.31). This sequential occurrence in time with relatively minor changes in form but continued improvement in structure of the skeleton (in particular, adaptations for feeding and locomotion) is regarded as anagenetic change (linear progression of evolution without branching into different lineages) rather than phylogenetic change, the latter

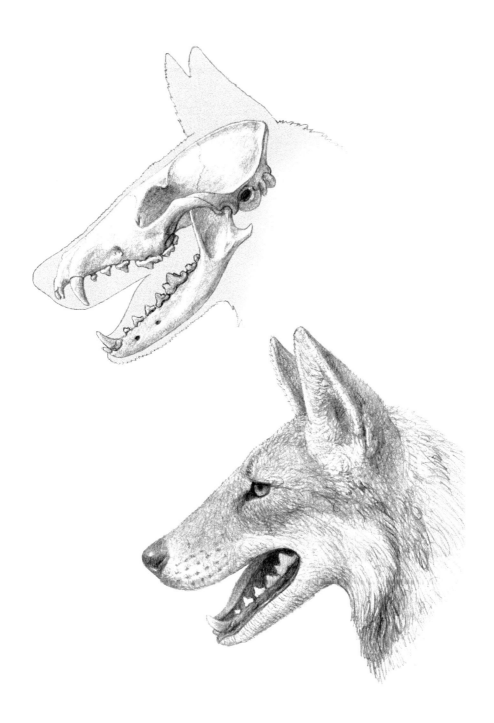

FIGURE 3.30

Eucyon zhoui

Skull and reconstructed head of *Eucyon zhoui* (4 Ma). Total length of skull: 17 cm.

FIGURE 3.31
Coyote
The extant coyote (*Canis latrans*). Shoulder height: 50 cm.

being accompanied by lineage splitting, which led to greater diversity of form. Fossils of a larger but still coyote-like species, *C. edwardii*, have been found in the later Pliocene record, along with *C. lepophagus*, in the southwestern United States. They may signify a phylogenetic event nearer the time of the former's appearance.

During this time, the first major immigrant species in this genus, *Canis armbrusteri*, entered the North American record. It was a wolflike form, larger than any native *Canis* species of the time. Large species of similar character are known in the fossil record of northern China, and they may have contributed such an immigrant to North America. *Canis armbrusteri* persisted in North America until the later Pleistocene, but in the middle part of that span (about 0.5 Ma), it gave rise in the midcontinent to the dire wolf (*C. dirus*), the largest species known within the genus (figures 3.32 and 3.33).

Typical of large *Canis* species, *C. dirus* has a hypercarnivorous dentition in which the shearing contact between the upper and lower teeth is emphasized at the expense of the grinding functions present in most canids. Just prior to the appearance of *C. dirus*, the continent was invaded by large and even more hypercarnivorous species of the Old World genus *Xenocyon*, which played an important role in the origin of the living dhole (*Cuon*) of Asia (figure 3.34) and the hunting dog (*Lycaon*) of Africa (figure 3.35). In North America, *Xenocyon* species were always rare, although equal in size to the native *C. dirus*. These immigrants were presumably no match in competition with the dire wolf because they had only a short existence in North America.

FIGURE 3.32
Dire wolf
Life reconstruction of the dire wolf (*Canis dirus*) (20,000 years ago). Reconstructed shoulder height: 77 cm.

The dire wolf spread into South America and generally across North America south of the Pleistocene glaciers blanketing the north of the continent. It may have become a llama specialist in its early evolution in North America and followed these camels into South America. Llamas in the past were more abundant in lowland pampas or alpine grassland environments of more temperate southernmost South America (as they are today). This region is where most dire wolf remains are found. Elsewhere in the continent, large cats and the more wolflike cerdocyonine canids would have competed with them.

After the end of the most recent glacial period (the past 30,000 years), a major climate change marked by glacial retreat and warming of high latitudes took place across the world. It was extensive enough to melt the glacial barrier across northern Canada, allowing mammals that inhabited arctic latitudes to extend their ranges to the south. This event changed the composition of the midlatitude mammalian fauna of North America, which was then invaded by Eurasian species. Along with the elk, caribou, mountain sheep, goats, and bison came their major predator, the gray wolf (*Canis lupus*) (figure 3.36). Most of these animals, including the wolf, had been inhabitants of arctic North America for much of the previous million years. All had undergone their major evolution in the Arctic, which then included Eurasia, eastward through Beringia, and across northernmost Canada into the Canadian arctic islands and perhaps into Greenland, a nearly globe-girdling range. Preferring arctic environ-

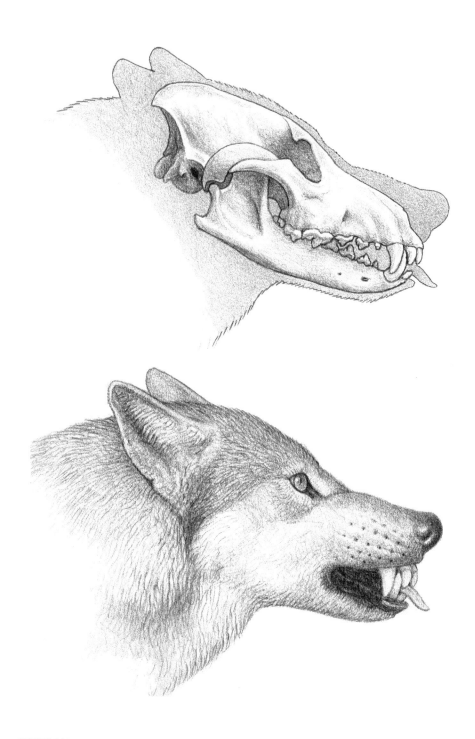

FIGURE 3.33
Dire wolf
Skull and reconstructed head of the dire wolf (*Canis dirus*). Total length of skull: 30 cm.

FIGURE 3.34
Dhole

The extant Asian dhole (*Cuon alpinus*). Shoulder height: 50 cm.

FIGURE 3.35
African hunting dog

The extant African hunting dog (*Lycaon pictus*). Shoulder height: 75 cm.

FIGURE 3.36
Gray wolf
The extant gray wolf (*Canis lupus*). Shoulder height: 75 cm.

ments, this fauna had wandered south very little in the previous glacial cycles, but conditions were finally favorable for an invasion of midcontinental North America.

The consequence of this invasion to canid history was significant. The northern immigrants included the red fox (*Vulpes vulpes*), the gray wolf (*Canis lupus*), and a large dhole species (*Cuon alpinus fossilis*), a surprising immigrant known from only a single site in the mountains of Gulf Coastal Mexico. Dholes (*C. alpinus*) still range into mountains along the southeastern margin of arctic Asia, hinting at the environments that enabled their spread into the New World (see figure 3.34).

The fossil record in the tar sands of Rancho la Brea in Los Angeles, California, shows the coexistence of the dire wolf and the gray wolf about 10,000 years ago; the dire wolf became extinct after that. Because the gray wolf did not live in South America (as far as we know), it was not a direct factor in the extinction of the dire wolf there, which seems to have been nearly coincident with its extinction in North America. The wolf's invasion of North America did not involve great disruption of the coyote's range; there was little overlap in predation because these species differ in size, socialization, and hence prey choice. Genetic work (Wayne and Jenks 1991) suggests that the coyote and the wolf may interbreed under certain conditions and produce viable offspring. Such interbreeding may have happened more frequently after the occupation of North America by Eurasian canids, which was devastating to eastern wolf populations, leading to their extinction or their merger with more successful coyotes. In Robert K. Wayne and S. M. Jenks's view, the red wolf (*Canis rufus*) of eastern North America may have arisen from coyote–wolf interbreeding

FIGURE 3.37
Red wolf
The extant red wolf (*Canis rufus*). Shoulder height: 65 cm.

along the boundary of the two species' range (figure 3.37). This boundary has been moving westward into the midcontinent after nearly three centuries of the constant culling of wolves. In this hypothesis, the red wolf is the product of a stress-initiated merging of two species whose lineages were separated geographically and temporally for more than 4 million years. A recent genetic reappraisal of populations of wolves from eastern Canada (Wilson et al. 2000) suggests that these animals (*C. lycaon*) represent another genotype not closely related to *C. lupus*, but perhaps to *C. rufus* and *C. latrans*. These animals thus formed a North American lineage separated from the true wolves (*C. lupus*), which invaded midcontinental North America at the end of the Pleistocene (0.01 Ma).

Is the Domestic Dog a Distinct Species?

By and large, domestic dogs have been treated either as a species of their own, *Canis familiaris*, or as a subspecies of the gray wolf, *C. lupus familiaris*. In 1758, the Swedish botanist and taxonomist Carol Linnaeus, founding father of modern biological classification, formally proposed the name *Canis familiaris* for the species and *Lupus* as a separate genus (now combined into *Canis lupus*) in his now famous volume *Systema naturae*. The Linnaean binomial classification carried no evolutionary connotation. The fact that Linnaeus placed the dog and the wolf in separate genera (*Canis* and *Lupus*, respectively) perhaps implies that he did not think of any connection between the two beside their morphological similarities.

The concept of species is a difficult issue fraught with intricate philosophical and practical problems. Theorists who keep track of such matters can count up to 20 different ways of defining species. Although we will not attempt to elaborate on all of them, we single out two popular definitions that serve to illustrate the difficulties we face in determining the classification of domestic dogs.

One of the most common views of species is the *biological species concept*, defined by renowned taxonomist and evolutionary theorist Ernest Mayr as "groups of interbreeding natural populations that are reproductively isolated from other such groups" (1969:26). As explained in many textbooks, this concept is predicated on the species barriers set up by reproductive isolation. In the case of domestic dogs, the question becomes whether or not dogs can interbreed with the gray wolf or, for that matter, with other species of *Canis* and, if so, whether their offspring would be viable. This is where the definition of *species* becomes messy because almost all dogs can and, given the opportunity, do occasionally interbreed with wolves. Therefore, a strict application of the reproductive criterion often leads to the conclusion that dogs can warrant no more than a subspecies recognition. Biologists who subscribe to the biological species concept tend to call domestic dogs *Canis lupus familiaris*, as designated by W. Christopher Wozencraft (1993) in the widely used compilation *Mammal Species of the World*.

A subspecies is equivalent to a race or a variety, and in mammalian taxonomy it usually corresponds to some kind of geographically distinct entity. For example, *Canis lupus occidentalis* refers to gray wolves that occupy western Canada and the northwestern United States, whereas *C. lupus lupus* is a race distributed in much of the middle latitudes of Eurasia. The number of geographic subspecies varies widely according to different specialists. E. Raymond Hall (1981) recognized as many as 24 North American subspecies of *Canis*, whereas Ronald M. Nowak (1979) recognized only five for the same region. Although all domestic dogs broadly overlap with gray wolves in their distribution, one can argue that dogs stay in human homes, whereas wolves stay in the wild—thus there is a "geographic" separation between dogs and wolves, somewhat similar to the separation of wild species by microhabitats. Critics who argue against a subspecific designation insist that reproductive barriers are not necessarily physical and that behavioral barriers (reluctance for populations to interbreed) are sufficient to mark distinct species. Such an argument brings us to a different way of looking at species.

An alternative view of species loosely called the *evolutionary species concept* allows the fact that when populations begin to diverge from each other, regardless of their reproductive potentials, they embark on different evolutionary trajectories. As long as two populations are on different evolutionary pathways leading to their respective descendant lineages, speciation has by definition occurred. Such a species concept has the advantage of doing away with such highly subjective and sometimes speculative measures as genetic or morphological differences (how much populations differ before we start to call them distinct species) and reproductive isolation (difficult and often impossible to ascertain, especially under natural conditions). In the case of dogs, one can argue that once they became associated with humans, they evolutionarily embarked on their own path, distinct from that of the wolves, and thus should be called a species of their own.

Raymond Coppinger and Lorna Coppinger (2001) argue that ecological separation among wolves, coyotes, jackals, and dogs is the key. Dogs have an obligatory symbiotic relationship with humans, a unique association that no other canines have. Yes, dogs occasionally breed with wolves, coyotes, and jackals, but their offspring are often intermediate in morphology and cannot compete with the "pure" forms; in other words, the hybrid is not really viable. In this sense, reproductive isolation is achieved at least in spirit, if not by strict biological barriers. (In this context, we need to keep in mind the distinction between *domestication* and *taming*. An animal is domesticated if it gives birth to offspring that are domestics, so the domestication is self-perpetuating. In contrast, taming occurs when wild animals are captured and then used for human purposes; in this situation, new captures are needed to replenish the tamed herds.) The Coppingers thus strongly argue that dogs are a distinct species, but they also point out that dogs represent a product of a "distinct evolutionary event," thus invoking, perhaps unwittingly, the evolutionary species concept.

If the biology/evolution controversy is not confusing enough, a further complication in the scientific name of dogs is that biological classification is intended to be a system that reflects some fundamental order of the natural world, as succinctly captured by Linnaeus's book title: *Systema naturae*. Most dog breeds, however, have been heavily selected (bred) by humans. This high degree of artificial selection did not occur in a "natural" world, and what we have created cannot be a natural product. Can we still use the biological classification for an artificial product? However, if dogs domesticated themselves (at least initially), as some argue (chapter 8), then the dog–human association is truly a natural result of symbiosis. In the latter case, a distinct species name might be warranted.

In evaluating all these options, one should keep in mind that biological classification often contains an element of judgment. Most taxonomists agree that classifications should be a close reflection of evolutionary relationships. But most of our notions of evolutionary relationships fall in the realm of historical hypotheses. If dog classifications reflect different ideas of the origin of dogs, then the controversies will justifiably continue. As vertebrate paleontologists, we prefer to use the subspecific designation *Canis lupus familiaris*, which captures the wolf–dog relationship. For other domestic animals, if the wild ancestors are known, the domestics are usually not given distinct species status; for example, in the case of the pig (*Sus scrofa*), the domestic pig is designated *Sus scrofa domestica*.

4 ANATOMY AND FUNCTION
HOW THE PARTS WORK

MUCH OF THE CANIDS' FOSSIL RECORD is preserved in the form of bones and teeth. Thus a great deal of anatomical information is lost during fossilization, a process that turns the bones and teeth into something similar in properties to the surrounding rocks and that destroys much of the soft tissues, including muscles, skin, and internal organs. However, fossil bones and teeth fortunately preserve a large amount of anatomical and biochemical information that is extremely valuable in determining an animal's anatomy, physiology, and, to a certain extent, behavior (figure 4.1). For example, limb bones, an essential component of all mammals' skeletons, can tell us much about the precise configuration of the limbs, which are important in deducing the animal's body weight and locomotion repertoire (the ways in which it walked, ran, climbed, or swam). Most skeleton bones also preserve traces of muscle attachments in the form of ridges, crests, and other superficial features that are the basis for the reconstruction of musculature in fossil animals. Starting with such solid information—the fossil evidence—paleontological artists are able to add layers of soft tissue over the skeletal frame in order to re-create an animal's appearance in life. All the illustrations of extinct mammals in this book were created using a detailed methodology based on skeletal morphology as the starting point for the restoration of unpreserved attributes (figure 4.2). Skull bones also preserve clues about the neural organs (the brain and nervous system) and the sensory organs (for hearing, smelling, and seeing). As part of the food ingestion system, skulls and jaws are also very informative regarding how food was procured and processed (figure 4.3).

When the dietary habits of an extinct carnivoran are studied, however, fossil teeth are the single greatest source of information. To gain an appreciation of the evolutionary history of the canids, one has to begin with some knowledge of the anatomical terms for teeth and bones, which are predominantly the stuff preserved in fossils.

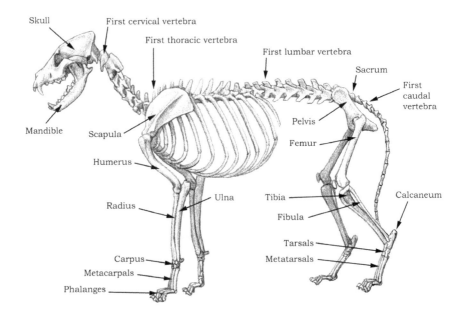

FIGURE 4.1
Gray wolf
Skeleton of the European gray wolf (*Canis lupus*), with labels showing key anatomical features.

Teeth

DENTAL FORMULA

Among carnivorans, canids have the most conservative (least-modified) teeth, with a bilateral complement of teeth divided into three upper and lower incisors, one upper and lower canine, four upper and lower premolars, two to three upper molars, and three lower molars. With the exception of supernumery teeth in animals such as the bat-eared fox (*Otocyon*), which grows an extra molar, and extreme cases such as dolphins, which have many more teeth than other mammals, a reduction in the number of teeth has occurred in mammals through evolution in various ways. Paleontologists have devised a system of dental formulas to express the number of teeth in shorthand, such as 3143/3143: the number to the left of the slash represents upper dentition (three incisors, one canine, four premolars, and three molars), and the number to the right denotes lower dentition. The numbers indicate only one side of the jaw, and the number for a full complement of teeth in the mouth must double the dental formula. With the exception of the bear family (Ursidae), the extinct bear dog family (Amphicyonidae), and the Canidae, all other modern families of carnivorans have had fewer teeth (that is, a reduced dental formula) over time,

FIGURE 4.2

Dire wolf

Step-by-step reconstruction of the dire wolf (*Canis dirus*), based on fossils from the late Pleistocene (20,000 years ago) of Rancho la Brea. *Top*, skeleton; *top center*, deep muscles; *bottom center*, superficial muscles; *bottom*, life appearance. The deep muscle layer is inferred largely from the direct observation of attachment areas on the bones. More superficial layers of tissue require a higher level of inference, and reference to modern related species is important.

with the cat family (Felidae) in the extreme end of the spectrum: 3121/3121. These reductions in the number of teeth are usually indications of specialized diets moving toward hypercarnivory. Another useful shorthand is the designation of an individual tooth by a combination of a letter—incisor (I or i), canine (C or c), premolar

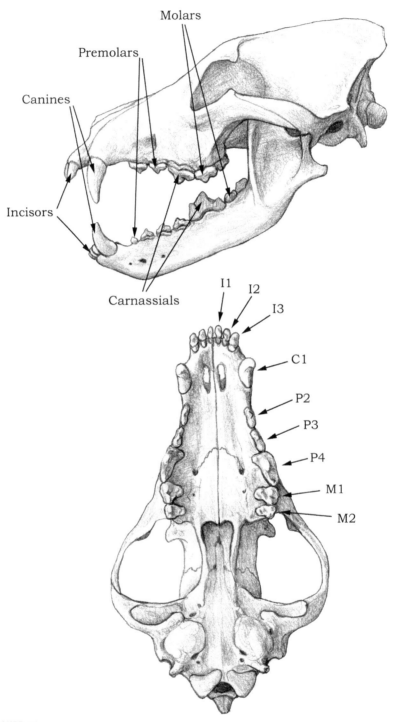

Molars

Premolars

Canines

Incisors

Carnassials

I1 I2

I3

C1

P2

P3

P4

M1

M2

FIGURE 4.3

Gray wolf

Top, skull and mandible of the gray wolf (*Canis lupus*) in lateral view; *bottom*, skull in palatal view, with the tooth positions labeled.

(P or p), and molar (M or m)—and a number (counted from front to back). Thus the entire dentition of a canid can be expressed as I1, I2, I3, C1, P1, P2, P3, P4, M1, and M2 in the upper jaw and i1, i2, i3, c1, p1, p2, p3, p4, m1, m2, and m3 in the lower jaw (see figure 4.3).

INCISORS

The incisors are the three small, neatly packed teeth with a flat cutting edge in front of the jaw, those that are clearly visible in front view when an animal opens its mouth. Incisors are rooted in the frontmost part of the jaws—in the premaxillary bone of the upper jaw and the dentary bone (frontmost tip) of the lower jaw. On close examination, there is usually one main cusp on the incisors and two smaller cusps flanking either side (one must make sure to look for these features in a relatively young individual because they are obliterated as the teeth are worn down in older individuals). One of the defining features of the borophagine canids is their possession of one to three small cusps on the side of the upper third incisor.

CANINES

The canines are the fanglike teeth that deliver the killing bites, and carnivorans display their canines for maximum effect in a threatening posture. All canids have prominent canine teeth—which fittingly bear the name of the dog family—in each jaw, and they are found in nearly all mammals (even humans have them, but they are reduced in size and look very much like incisors). The upper canines are technically located at the suture zone of the premaxillary and maxillary bones in the upper jaw and shear with the lower canines, which are slightly in front. In canids, the size of canine teeth tends to be somewhat variable between the sexes. Male canids of fox- to wolf-size species usually have canines that are 3 to 6 percent larger than the females' canines (Gittleman and Van Valkenburgh 1997).

PREMOLARS

The premolars are the teeth between the canines and the molars. In canids, with the exception of the upper fourth premolar (see "Carnassials"), the premolars are formed by a tall main cusp and one or two smaller accessory cusps in front of and behind the main cusp. As do the incisors and canines, the premolars consist of two sets: the deciduous (or milk) premolars and permanent premolars. The last deciduous premolars often look quite similar to the molars, but the permanent premolars adopt a radically different construction. Canid premolars tend to be stable in most lineages, but among taxa that evolved a bone-crushing adaptation, the premolars (particularly those of the rear end of the series, including the upper carnassials) can be substantially enlarged to become the bone-cracking teeth. Such an adaptation is most obvious in some lineages of the borophagines and, to a lesser extent, the hesperocyonines.

MOLARS

The molars (from Latin *mola* [mill, millstone]) are the backmost series of teeth. They differ from the premolars in their lack of a deciduous precursor; that is, permanent molars emerge without milk teeth having preceded them. In canids, with the exception of the lower first molar (see "Carnassials"), the molars are formed by a low platform with a series of low cusps closely aligned with each other on the crown surface. These small cusps occlude with each other, so the upper and lower molars function like a millstone. These grindstone-like teeth are commonly well developed in canids (but not as extremely developed as in ursids), and they permit flexibility in a canid diet that mixes meat with vegetable matter and insects. The relative size of the molars is an excellent indicator of the diversity of an animal's diet.

CARNASSIALS

A specific pair of teeth is very important in defining the order Carnivora: the carnassials, specifically the upper fourth premolar and the lower first molar, which function as the main teeth for cutting the muscles and tendons of killed prey. The carnassials are specially modified in all carnivorans to form a pair of shearing blades that act like scissors. This scissorlike carnassial pair was the most definitive adaptation of the ancestral carnivorans and has been passed down to all descendant lineages. Therefore, the order Carnivora is literally defined by its members' common possession of a pair of carnassial teeth. In canids, the carnassials are usually well developed enough to handle most of the meat-cutting requirements.

Hypercarnivory Versus Hypocarnivory

The concept of dental adaptation is very important. In 1956, Spanish vertebrate paleontologists Miguel Crusafont-Pairó and Jaime Truyols-Santonja published an insightful study of carnivoran functional morphology in which they categorized carnivorans into hypercarnivores and hypocarnivores based on their dental morphology.

A *hypercarnivore* is an animal that has elongated the shearing blade of the carnassial teeth at the expense of the grinding part of the dentition (usually molars). The most extreme example of a hypercarnivore is the cat, whose teeth essentially include only the shearing part of the dentition—a pair of long, thin-bladed carnassials—with the grinding part (molars) behind the carnassials being strongly reduced in size (figure 4.4). Such a hypercarnivorous adaptation is thought to be related to a diet that is made up almost exclusively of meat.

In contrast, a *hypocarnivore* is an animal that has shortened the shearing blade of the carnassial teeth and enlarged the grinding part of the dentition behind the carnassials. The most extreme example of a hypocarnivore is the bear, in which the shearing part of the carnassial teeth is radically reduced and the grinding parts of the molars are extremely broadened (figure 4.5). A hypocarnivore tends to have a far more varied diet that includes meat, insects, fruits, and roots. In the most extreme case of hypocarnivory, the giant panda (*Ailuropoda melanoleuca*), which is an ursid, has become a dedicated bamboo eater.

In between these two extremes, most of the carnivorans, including the majority of the canids, have teeth that are neither extremely hypercarnivorous nor particularly hypocarnivorous. We call these intermediate forms *mesocarnivores* (from Latin *meso* [middle]). In this book, when we discuss hypercarnivorous canids, we refer to canids that have relatively more elongated dental shearing blades and at the same time a relatively reduced grinding part of the dentition, even though these canids are by no means in the same league as the cats in terms of dental specialization (figure 4.6). Likewise, we refer to canids in which the opposite trend has occurred as hypocarnivorous (figure 4.7). But here again, canid hypocarnivory is by no means as extreme as ursid hypocarnivory.

Skulls, Jaws, and Teeth as Indicators of Diet

An important fact of vertebrate paleontology is that dental morphology is highly sensitive to a mammal's diet. Such a close association between the shape of teeth and the diversity of diet permits paleontologists to learn much about what extinct animals ate. The teeth of canids have changed flexibly toward either hypercarnivory or hypocarnivory during their evolutionary history, which is probably one of the most important reasons for their success as a group in adapting to fast-changing environments. Because of their relatively conservative, unspecialized dentition, ancestral canids (such as *Hesperocyon*, *Archaeocyon*, and *Leptocyon* [chapter 3]) were apparently free to evolve toward either hypercarnivory or hypocarnivory, capitalizing on new opportunities that arose from time to time.

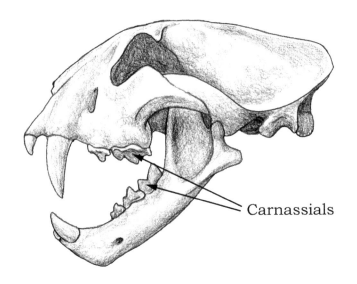

Carnassials

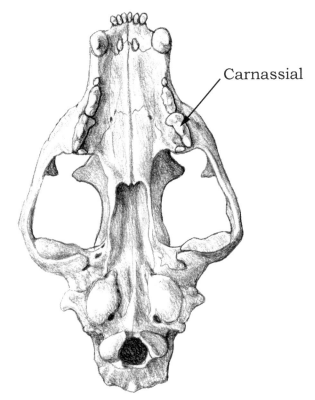

Carnassial

FIGURE 4.4

Leopard

Skull of a typical hypercarnivore, the leopard (*Panthera pardus*; family Felidae). The cheek teeth in front of and behind the carnassials are either reduced or lost, and the carnassials themselves are narrow, bladelike teeth.

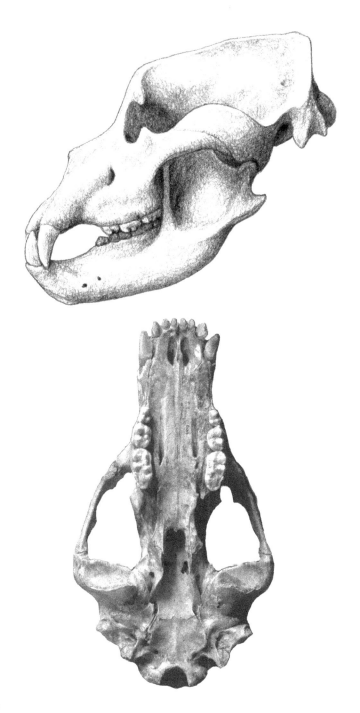

FIGURE 4.5

Cave bear

Skull of a typical hypocarnivore, the cave bear (*Ursus spelaeus*; family Ursidae). The premolars are reduced or lost, the molars are large and broad, and the carnassials are broadened.

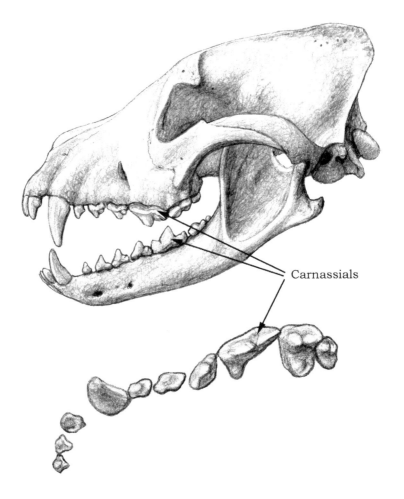

Carnassials

FIGURE 4.6

African hunting dog

Top, skull and mandible of a hypercarnivorous dog, the extant African hunting dog (*Lycaon pictus*); *bottom*, detailed occlusal view of the upper-right dentition. Notice the long, bladelike carnassials.

The vertebrate skull serves the all-important function of protecting the brain. However, if protection of the brain were the skull's only function, then there would be no need for variety in skull shape because a rounded, hard-shelled braincase would be all that any vertebrate needed to shield the brain adequately against common, self-inflicted mechanical damage (as opposed to damage inflicted by predators, which is a different matter). The fact that vertebrate skulls do vary substantially from group to group implies that some other factor is also at play. That factor is often related to food processing.

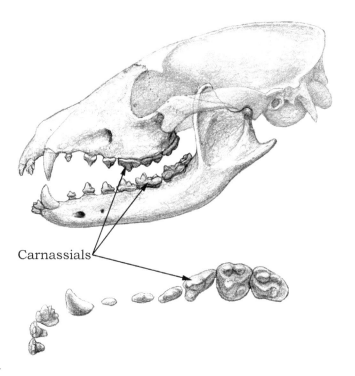

Carnassials

FIGURE 4.7

Cynarctus

Top, skull and mandible of the hypocarnivorous canid *Cynarctus* (10 Ma); *bottom*, detailed occlusal view of the upper-right dentition. Notice the relatively small carnassials and the enlarged, broad molars, as well as the extra cusps along the side of the upper third incisor.

As the location of initial food ingestion and mechanical processing, the skull and lower jaw are perhaps the most critical devices in meeting the demands of food intake. For carnivorans, the skull and jaws also aid in the all-important function of killing prey, and it is this function that often defines the overall skull shape in carnivorans. One of the most variable and, by implication, evolutionarily most selected features is the length of the muzzle, or *rostrum*. It is easy to see differences in this feature among cats and dogs. In general, cats have a much shorter rostrum than do dogs, and this evolutionary divergence can be traced to the very beginning of both families.

LENGTH OF ROSTRUM

Over time from their viverravid beginning, feliform carnivorans have evolved a progressively shorter rostrum. This shortening is associated with the reduction and loss of the anterior premolars (usually first and second premolars) and of the posterior molars (usually second and third molars). The net effect of this shortened rostrum is more posteriorly positioned incisors, canines, and premolars. With the movement of these teeth backward, a more powerful bite was achieved with the same muscles. Such an arrangement is not difficult to understand. When we humans try to crack a hard nut, for example, we tend to place it farther back in our mouths toward the molars, where our temporalis and masseter muscles (the main muscles for closing the mouth) can exert their maximum strength.

The biomechanical principle behind this arrangement is a simple leverage system. Mammalian jaws are anchored on the posterior tip, and this anchoring point (the *mandibular condyle*) acts as a hinge around which the jaws rotate (as in the common phrase "His jaw dropped") (figure 4.8). The temporalis and masseter muscles, the main musculature for closing the jaws, are inserted in and around a deeply pocketed area called the *masseteric fossa* on the ascending ramus (vertical beam rising behind the last molar) of the jaws. The masseteric fossa is always in front of (anterior to) the hinge point, and contractions of the temporalis and masseter muscles acting on the masseteric fossa pull the jaws upward and forward. In such a leverage system, everything else being equal, the closer a food item is to the mandibular condyle (that is, the more posteriorly located the item is), the more powerful a bite on the item would be. Because the killing bite by a carnivoran is almost always delivered by the canine teeth, which are located toward the front tip of the upper and lower jaws, the shortening of jaws can be an effective mechanism in bringing the canines closer to the mandibular condyle, thereby increasing the power of the bite.

As noted, caniform carnivorans have remained conservative in their cranial morphology from the very beginning. Canids inherited this conservative plan and so started out with an unreduced set of teeth. To accommodate the full dentition, the rostrum is relatively long, with the incisors and canines positioned more forward. From this ancestral condition of a relatively long rostrum, canids were flexible enough to be able to evolve toward either a shortened rostrum or a more lengthened rostrum. Hypercarnivorous forms tended to acquire a shorter rostrum, paralleling the condition in cats, and a short rostrum can be seen in several groups of the hesperocyonines and borophagines, although neither ever achieved the extremely shortened jaws seen in the felids (figure 4.9). In the subfamily Caninae, however, a lengthening of the rostrum is evident from the very beginning of the lineage (*Leptocyon*), and this feature was passed down to all its descendants. A long rostrum increases the area of the nasal cavity (see "Turbinates") and presumably enhances olfactory functions. Dogs are thus more smell oriented than cats, which are more vision oriented. A lengthening of the rostrum is sometimes also found in mammals that are ant or termite eaters because they use the added length of the rostrum and tongue to reach into narrow spaces. We may speculate about the food habits of early

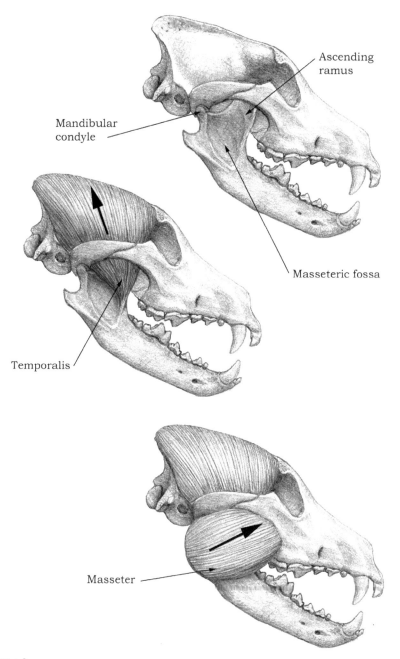

Ascending ramus

Mandibular condyle

Masseteric fossa

Temporalis

Masseter

FIGURE 4.8

Gray wolf

Skull, mandible, and principal jaw-adductor muscles of the gray wolf (*Canis lupus*). *Top*, skull and mandible showing the position of the ascending ramus of the mandible, the masseteric fossa, and the mandibular condyle, which is the hinge around which the jaw rotates; *center*, the temporalis muscle attaches to the top and anterior margin of the ascending ramus of the mandible, and the main direction of its pull (large arrow) is upward and backward; *bottom*, the insertion of the masseter muscle occupies much of the masseteric fossa, and the main direction of its pull (large arrow) is upward and forward. The force transmitted to the teeth by the contraction of these muscles is greater if the teeth are located closer to the hinge of rotation. Consequently, the shorter a carnivore's muzzle, the greater the force exerted at the canine tips during the bite.

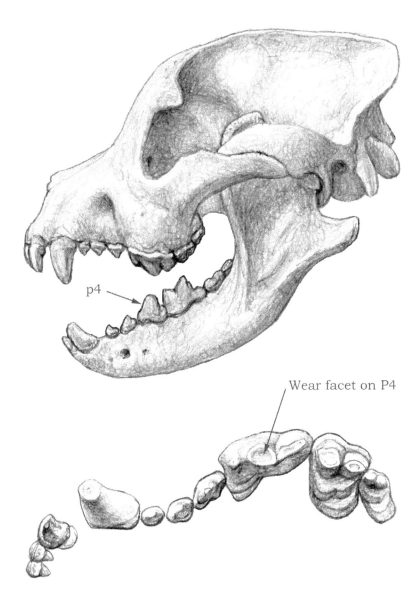

FIGURE 4.9

Borophagus

Skull and dentition of a bone-cracking dog, *Borophagus*. Notice the shortened rostrum, domed forehead, strengthened mandible, and robust teeth, especially the lower fourth premolar (p4). The upper carnassial (P4) shows horizontal wear associated with bone cracking.

canine progenitors that may have been more insectivorous in their diet, but so far there is little evidence to confirm such speculation.

STRENGTH OF JAW

The strength (or the cross-section area) of the jaws is another important factor in their mechanical capabilities. A shorter jaw is generally optimally suited for delivering a powerful bite. Although canid jaws have probably never been as powerful as felid jaws, the less powerful jaws in canids must have been more or less adequate as killing devices, judging from the canids' success over a very long period of time. However, for those lineages that evolved bone-cracking dentitions, the jaws had to be strengthened in order to cope with the added stress. Thus it is in these lineages—such as *Enhydrocyon*, *Aelurodon*, and *Borophagus*—that the jaws became more massive, with a thickened horizontal ramus (see figure 4.9).

DOME OF HEAD AND FRONTAL SINUS

Hypercarnivorous forms, such as felids, generally have a shortened rostrum and a rather arched skull top as a result of an inflated forehead area. On a lateral (side) view, this arching can be seen as an upward swell of the skullcap above the eyes. Such a domed head may offer some advantages in its reinforcement of the bony structures that anchor the canine teeth. It is seen in all felids and, to a lesser extent, in hyaenids (except for the ant-eating aardwolf [*Proteles cristatus*]).

Large, hypercarnivorous canid species also tend to have enlarged foreheads. This shape is accomplished by inflation of the frontal sinuses, or the air spaces beneath the frontal bones. Through the expansion of the frontal sinuses, particularly by the upward swelling of the frontal bones, the head often displays a prominent dome in profile view. In extreme cases, the frontal sinus is extended backward toward the back tip (inion) of the skull, as seen in *Borophagus diversidens*, and, as a result, a large air space exists between the braincase and the parietal bones above it. Swedish vertebrate paleontologist Lars Werdelin (1989) has postulated that such a domed head in bone-cracking canids and hyaenids serves to transmit the great stresses of the premolars to the back of the skull, thereby reducing the bending stress of the rostrum (see figure 4.9).

Ear Bones

Bones in the ear region are the most intricate of the internal skeletal systems in mammals. Modern carnivorans typically have a hard-shelled, rounded bony housing in the head called the *tympanic bulla*, which, together with the tympanic membrane, holds three tiny ear ossicles (malleus, incus, and stapes) inside. Such a fully ossified enclosure of the middle-ear space is lacking only in the highly conservative

African palm civet (*Nandinia*), which is a basal feliform distantly ancestral to all living catlike carnivorans, such as felids and hyaenids. The tympanic bulla in the palm civet is still partially formed by soft, cartilaginous tissue. A cartilaginous bulla is apparently the ancestral condition for all carnivorans. Indeed, all archaic carnivorans, such as miacids, lacked a bony bulla, and their fossil representatives always have a naked ear region without bony cover (it is possible that some of the early carnivorans may have had partially ossified bullae, but the bullae are not preserved in the fossil record because they were easily separated from the skull or destroyed during fossilization).

Canids were the first among living families of carnivorans to evolve an ossified bulla. From the very beginning, *Prohesperocyon* sported a hard-shelled tympanic bulla. Furthermore, all canid bullae are partially partitioned by a ridgelike structure called a *semiseptum*, which seems to work as a strut to reinforce the structural integrity of the bulla rather than to divide the bulla space. Felids, in contrast, have a full septum that divides the bulla into front and back chambers. Whatever the function of the septum, such a structural difference between canids and felids is a convenient way to distinguish these two families when the ear region is preserved in fossil materials (figure 4.10).

A rigid bony housing for the middle-ear space not only protects the delicate ear ossicles in their intricate linkage for sound transmission, but also maintains a fixed volume of air undisturbed by jaw movements (figure 4.11). Some jaw muscles are attached to the bulla, and when in motion, they would deform the bulla if it were flexibly constructed of cartilage; such a disturbance is probably not desirable because of its interference with hearing. The size of the bulla housing and thus the volume of air inside are optimized toward the hearing characteristics of a particular animal. A conspicuously enlarged bulla is present in the fennec fox (*Vulpes zerda*), of the Sahara and Arabian deserts, which also possesses a greatly enlarged external ear (pinna). A larger volume of air in the bulla appears to be related to the enhancement of low-frequency hearing (just as a larger audio speaker can resonate better for base notes). Enhanced low-frequency hearing in an open, desert environment has apparently been of adaptive value to the fennec fox.

Otarocyon, a basal borophagine canid from the early Oligocene (34 Ma) of western North America, also developed a strikingly large bulla for its small size (chapter 3; see figure 3.13). This extraordinary structure poses a tantalizing question: Why was such a big ear developed in the beginning of canid evolution? Was it in response to a more open, grassy landscape in the early Oligocene, as opposed to a closed, wooded environment in the Eocene (55 to 35 Ma)? Alternatively, could *Otarocyon* have hunted prey such as rodents and invertebrates in burrows by listening to the low-frequency sounds transmitted through the earth?

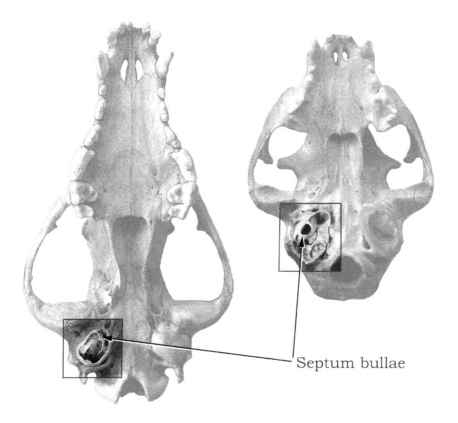

Septum bullae

FIGURE 4.10

Comparison of canid and felid skulls

Ventral views of a canid skull (*left*) and a felid skull (*right*), with the auditory region highlighted. The bulla has been dissected to show the inner structure, with the dividing wall, or septum.

Turbinates

Modern dogs are famously known for their acute sense of smell; people marvel at rescue dogs and at the ability of sniff dogs to detect trace chemicals that are completely beyond detection by humans. What is not well known by the general public, however, is that the canid nasal cavity features an elaborate set of bones called *turbinates*, which are associated with odor detection and breathing (figure 4.12). Turbinate bones are delicate, paper-thin scrolls of bony tissue that are outgrowths of the inner walls of the nasal passage (including the maxillary, nasal, and ethmoid bones). The ethmoturbinates are covered mostly by olfactory epithelium, or thin layers of skinlike tissue exposed to the air passageway and packed full of olfactory receptor neurons that transmit the sensing of an odor to the olfactory bulb of the forebrain. Heightened olfactory sensitivity is actually a characteristic of all mammals, and thus

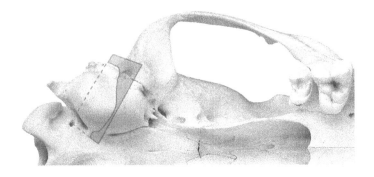

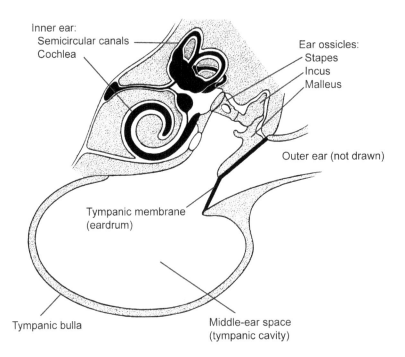

Inner ear:
Semicircular canals
Cochlea

Ear ossicles:
Stapes
Incus
Malleus

Outer ear (not drawn)

Tympanic membrane
(eardrum)

Tympanic bulla

Middle-ear space
(tympanic cavity)

FIGURE 4.11

Anatomy of a canid ear

Ventral view of a coyote skull showing right auditory bulla (*top*); the rectangle across the bulla indicates the location of the transverse cross sectional view (*bottom*). The cross section (modified after Evans and Christensen [1979:fig. 19.3]) shows the relationship of the outer-, middle-, and inner-ear regions and three ear ossicles.

the keen sense of smell in canids is inherited from its mammalian ancestors (the relatively obtuse sense of smell in humans is probably a reflection of our increased reliance on our optical senses, such as color and stereo vision).

Maxilloturbinates and nasoturbinates are covered by respiratory epithelium, a vascular mucous membrane over which the inspired air passes. These turbinates function mainly to warm and moisten the air during inspiration and to conserve

Turbinates

FIGURE 4.12
Red fox
Frontal view of the skull of an extant red fox (*Vulpes vulpes*), showing the turbinates inside the nasal cavity.

heat and water during expiration. On inspiration, the relatively cool and dry incoming air is warmed and saturated with water moisture as it passes over the moist, warm vascular maxilloturbinates. On expiration, the process reverses: warm and saturated air passes back through the just-cooled maxilloturbinates, helping to condense water onto their surface and thereby conserving water for the next breathing cycle. Some studies have suggested that as much as 80 percent of the heat and moisture are retained in the turbinates during expiration through the nose. In canids, the maxilloturbinates, the principal site for heat and moisture exchange, are greatly expanded. The complex and delicate scrolls of maxilloturbinates create an increased surface area within a confined space inside the nasal cavity, thereby increasing the efficiency of heat and water exchange. Such a complex of turbinates may have adaptive values for arid or cold environments, or both, although the size of the turbinates in canids does not strictly reflect the temperature and aridity of their environments. It is nonetheless interesting to speculate on the advantage of an elaborate maxilloturbinate system in arctic canids, such as arctic wolves and foxes, which have the widest distribution among any carnivorans in the vast frozen terrain in northern Eurasia and North America. The arctic faunal realm was an important component in canine evolution in the late Cenozoic (3 Ma) (chapter 6), and canines' turbinates in this era may have played a key role in their survival in this harsh environment.

The complexity of canid turbinates is demonstrated in a third purpose for them: the cooling of the blood in the brain. Heat exchange on the maxilloturbinate surfaces cools the blood that supplies the brain via the heart, a process similar to panting in hot weather. In contrast, felids have a countercurrent heat-exchange mechanism that serves to cool arterial blood by immersing arteries within a venous sac called the *rete mirabile*, which is hidden behind the eyeballs. Maintaining a brain temperature cooler than the core body temperature during high physical exertion is important to all warm-blooded mammals that live in hot climates. Canids are capable of prolonged chases, in contrast to the common ambush predation of felids, and complex turbinates may well have played an important role in enabling this predatory strategy.

Canid maxilloturbinates thus appear to be valuable in both cold and hot climates, which may explain why the sizes of the turbinates in different species of canid do not strictly correlate with the environmental temperature. A large, complex maxilloturbinate system may be adaptive in either situation. It also explains canids' long noses in contrast to felids' short noses.

Bone Cracking

Although most of us tend to dismiss the bones on our dinner table as nothing more than a hindrance in our consumption of meat, bones are a source of high-value nutrition in the form of marrow. The center of a hollow long bone (usually various limb bones) is filled with yellow marrow, a fatty substance rich in protein and monounsaturated fats (many cultures use bones for flavoring in cooking because of this internal content). However, the bones themselves also contain an elaborate meshwork of calcium phosphate, and this same sturdy skeleton, built to withstand powerful exertion, is extremely tough to crack. Consequently, not all carnivorans can make use of bone marrow when they consume their prey.

To get at the nutritious marrow, one has to crack open the long bones, which commonly are the hardest bones in the entire skeleton because they bear an animal's full weight. With the exception of some sea otters that use stones to crack shells, most carnivorans have to crack such bones with their own teeth if they desire what is inside the bones.

Teeth are the hardest biological materials produced in mammals because of the white shiny substance, called *enamel*, that coats the main body of the tooth (the crown). Made of a crystalline form of calcium phosphate, enamel is harder than bone. However, having a harder substance on the teeth is not enough to handle the bones. Teeth must be sufficiently massive to avoid being shattered during feeding. They also must be situated at the right location to deliver maximum pressure, and there must be a suitable musculature and jaw architecture to withstand the massive strain generated during bone cracking. The best example of such a bone-cracking adaptation is seen in the modern spotted hyena (*Crocuta crocuta*) of the African savannas. The animal has a massive jaw fitted with extremely robust premolars whose crowns are shaped like a short bullet up to 0.5 inch in diameter (figure 4.13). Because of the reduced number of molars in hyaenids, the premolars are located farther back to allow the best leverage during a bite. Feeding by modern hyenas is perhaps the most efficient of any carnivorans. A pack of hungry hyenas can quickly consume a carcass in a matter of minutes, leaving nothing behind, including the bones. The crushed bones will go through the hyenas' digestive system and ultimately be excreted as small fragments.

To strengthen the bone-cracking teeth, the enamel's microstructure is aligned to form a highly elaborate system of Hunter-Schreger Bands (HSB). The enamel in the HSB is woven in a pattern of crisscrossing crystalline fibers that stop cracks when they begin to develop due to high stress, much like a densely woven cloth can resist stress in all directions. Under high magnification (such as using a scanning electronic microscope), the enamel in bone-cracking teeth can be seen to form a zigzag pattern that maximizes their strength (figure 4.14).

During most of the canids' evolutionary history, they were confined to the North American continent (chapter 7), where hyaenids were not present (only a single genus of hyaenids, *Chasmaporthetes*, managed to immigrate into North American in the Pliocene [around 5 Ma]). Several lineages of hesperocyonines and borophagines began to develop teeth and jaws capable of cracking bones. Some of them—such as *Enhydrocyon*, *Ectopocynus*, and *Aelurodon*—had multiple heavy premolars on the jaw, like the hyaenids. Others, such as *Epicyon* and *Borophagus*, developed a massive lower fourth premolar that was far taller and larger than the preceding premolars. This latter strategy of concentrating stress delivery on a single locus was apparently a highly successful one and gave rise to a long and diverse *Epicyon–Borophagus* lineage.

Standing Posture

Canids, like hyaenids and some felids, are often pursuit predators, securing food by persistent chasing of prey. That canids are good runners is well known, as seen in greyhound and dog sledge races. Typical of all good runners, canids have long, graceful legs. The advantage of a long leg is obvious: everything else being equal, the longer the legs, the longer the stride. The length of a stride is the distance between steps, and, given the same frequency of stride, a longer stride naturally leads to a faster speed. It is thus almost universally true that the leg length of all fast runners (predator and prey alike) tended to increase in evolution.

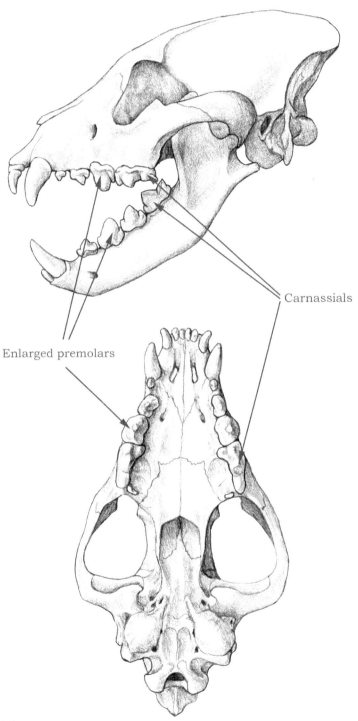

Carnassials

Enlarged premolars

FIGURE 4.13

Spotted hyena

Skull of the extant spotted hyena (*Crocuta crocuta*). Notice the enlarged, robust premolars, which are placed far back in the jaws thanks to the space gained with the loss of the posterior molars.

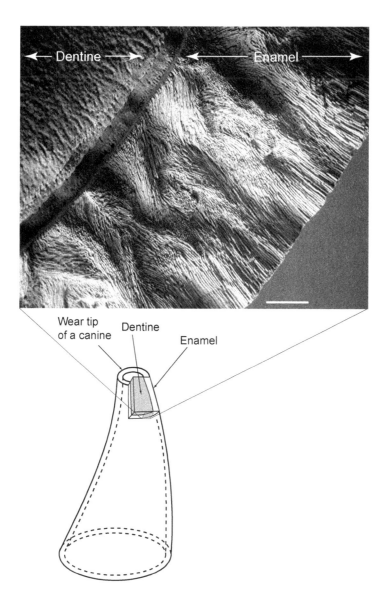

FIGURE 4.14

Enamel on the tooth of *Borophagus secundus*

Enamel microstructure of an upper canine tooth of *Borophagus secundus* under scanning electronic microscope. Individual "fibers" in the enamel band are enamel crystallites that are woven in intricate ways (HSB) to strengthen the tooth. The image is taken on a cross section shown in the lower diagram. The white scale bar on the lower right corner equals 100 μm. (Image courtesy of John M. Rensberger)

To maximize the benefits of a lengthened leg, the lengthening is almost always concentrated in the distal segments of the limbs (those segments toward the finger and toe tips), rather than the proximal segments (those segments close to the body). For example, the tibia (shin bone) is proportionally more elongated than the femur (thigh bone). The reason for this disproportionate length is that the distal segments have less muscle mass attached to them, and the heavy, lifting muscles are concentrated mostly in the proximal segments. Less muscle means a lighter limb in the distal end, which reduces the angular momentum and thereby eases the muscle exertion required for swinging the limbs.

The adoption of an increasingly more erect posture also increases stride length. Broadly speaking, there are three categories of standing posture: *plantigrady*, *digitigrady*, and *unguligrady*. In a plantigrade posture, the animal walks on the entire palm and sole, and the main joint that marks the beginning of the foot is the ankle joint, as in humans and bears. In a digitigrade posture, however, the animal walks on the fingers and toes, and the palm and heel parts of the hands and feet are lifted into the air. Thus the main joint is the joint between the base of the finger/toe bones and the metapodials just before the former. Among carnivorans, the canids, felids, and hyaenids have adopted a digitigrade posture (figure 4.15). The immediate benefit of such a posture is that a segment of the proximal hand and foot (the metacarpal and metatarsal, respectively) is converted into a segment that contributes to the stride. In an unguligrade posture, the animal walks on the very tip of the fingers and toes (distal phalanges), thus lifting up an additional two segments of the digits (the proximal and medial phalanges). The word *unguligrade* derives from *ungulate*, which refers to all hoofed animals that use their terminal phalanges for walking and running, including such well-known examples as horses, sheep, and cows. Thus in the unguligrade posture, the conversion of limb segments is literally and figuratively pushed to the extreme, as in the case of a ballet dancer. The best example of an unguligrade is the horse: its hooves are equivalent to humans' terminal fingers and toes with nails; the horse stands literally on the very tips of its middle "finger" and "toe."

In general, digitigrade animals' hands and feet can be recognized by closely packed or narrower metacarpals and metatarsals, in contrast to the more spread-out appearance of plantigrade animals' hands and feet. Thus in the former, the metacarpals and metatarsals are nearly parallel to each other, and the distal ends of these bones are nearly touching each other.

The benefit of a more erect posture for runners is fairly obvious: stride length is instantly gained without additional anatomical complexity. But if such a benefit is universal, why didn't canids (or, for that matter, felids and hyaenids) become unguligrade, as did almost all their prey? Doesn't a digitigrade or plantigrade posture place a disadvantage on the predators, which have to settle for shorter legs? The answer appears to lie in the fact that carnivorans had to balance their need to increase speed with their need to use their claws as tools for manipulating the prey. A horse's hooves are nothing but a pad to cushion its steps, and a horse does not have to use

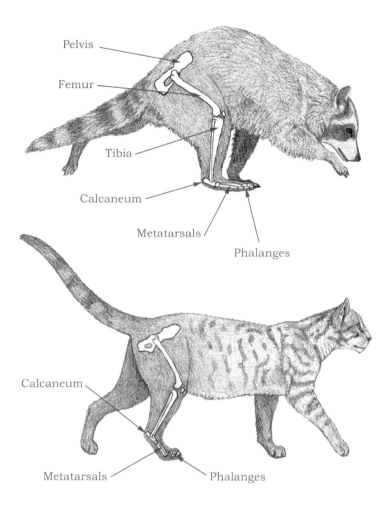

Pelvis

Femur

Tibia

Calcaneum

Metatarsals

Phalanges

Calcaneum

Metatarsals

Phalanges

FIGURE 4.15

Comparison of plantigrade and digitigrade posture

Comparison between a plantigrade carnivore, the raccoon (*top*), and a digitigrade carnivore, the house cat (*bottom*). Notice that when the foot carries the weight of the body, the calcaneum, or heel bone, is high above the ground for the cat, but close to the ground for the raccoon.

its hooves other than for walking and running. Carnivorans, in contrast, use their claws (equivalent to a horse's hooves) for manipulating, digging, climbing, and even killing.

Another important development in fast runners is the loss of peripheral digits. The digits that are not central to bearing weight were reduced and ultimately lost in all ungulates through much of their evolutionary history. Again, a horse serves as a good example. Modern horses have lost all but the single central digit (the middle finger and toe).

From the beginning, canids (such as *Hesperocyon*) had already acquired a fairly advanced hand and foot, with closely appressed fingers and toes, indicating a semi-digitigrade posture. But *Hesperocyon* retained a relatively well developed first digit (thumb and big toe) and a relatively deep claw. The first digits in later canids were progressively reduced, and the claws became smaller. All modern canids show the loss of a functional first digit, and their hands and feet have four fingers and toes.

In summary, all runners have a combination of lengthened distal limbs, more erect posture, lost lateral digits, and reduced muscle weight in the distal segments. As herbivores, ungulates can often push all these components to the extreme, whereas carnivorans must balance the demand for speed with the need for manipulation of food.

Claws

Claws generally serve three important functions in carnivorans: to help secure prey, to grasp trees when climbing, and, for some, to dig holes in the ground. Recent studies show that ancestral carnivorans were arboreal in the early Cenozoic (around 55 Ma). They had relatively short, strong arms and legs for climbing up and down trees, and flexible wrist and ankle joints for greater degree of rotation of their hands and feet—all necessary for a tree dweller. Often associated with arboreality are deep (in a side view) and sharp claws that can sink into tree bark for a firm grasp. Such sharp claws also come in handy for carnivorans that have to tackle prey of large size.

Claws are formed by a horny sheath (made of the same rigid material as the fingernails of humans) and have bony support beneath the terminal phalanx, or the last segment of the finger bones. The horny material consists of keratin, a colorless protein of the same origin as skin, hair, and horn sheathes in all mammals. Although claws are strong enough to resemble hard tissues such as bones, almost no claw is preserved in fossils because the keratin proteins usually decompose quickly after death. Thus we can only deduce claw shapes from the shape of the underlying bony terminal phalanges (figure 4.16).

One of the most important properties of claws is that they can be retracted into the back of the hands and feet. Retractile claws are an efficient way to preserve the sharpness of the tips while an animal is walking and running on land. Almost all felids have retractile claws that are pulled back when the felids are moving on the ground. They extend their claws when climbing trees or securing prey. The mechanisms that permit a retractile claw are a combination of an elastic lateral dorsal ligament, which passively pulls the claws backward toward the medial phalanges (the segment of finger bone immediately behind the terminal phalanges), and a deep dent on the dorsal surface of the medial phalangeal bone, which accommodates the

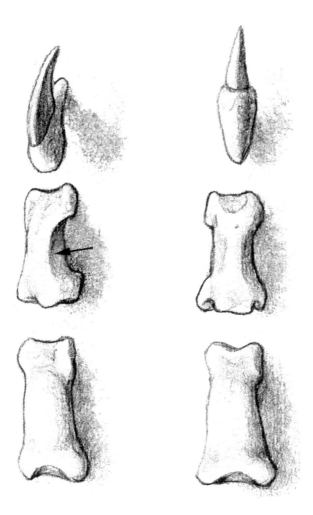

FIGURE 4.16
Comparison of felid and canid phalanges

Phalanges of the central digit of the right forefoot in a cat (*left*) and in a dog (*right*), shown disarticulated in dorsal view. The first phalanges are at the bottom of the figure, and the third ones (claw phalanges) are at the top. Notice that the dog's claw is less compressed laterally and that the cat's second phalanx has a marked concavity in the lateral margin (marked by an arrow), which is where the third phalanx fits during retraction of the claw. The dog's second phalanx, in contrast, is almost completely symmetrical.

pulled-back claws. This dent creates an asymmetrical appearance on the medial phalanges when they are looked at from above. Fortunately, this asymmetry is readily apparent in fossil foot bones, allowing paleontologists a ready means to distinguish between retractile and nonretractile claws.

As noted, a retractile claw originally evolved in ancestral carnivorans as a function of tree dwelling. As later carnivorans quickly discovered, sharp claws are also potent weapons in securing their prey. For example, a modern lion can ride on top of a wildebeest, rodeo style, using its sharp claws to grab hold of its prey's skin, while trying to deliver a killing bite on the neck. This ability allows a felid to secure its prey by itself. In contrast, canids are unable to cling to their prey with their blunt claws.

In an interesting twist of evolutionary history, canids apparently failed to develop retractile claws early in their history. As they became increasingly more cursorial through the lengthening of their limbs and the adoption of a more erect posture, they lost the ability to retract their claws. According to the fossil record, this change occurred shortly after *Hesperocyon*, which still possessed a moderately deep claw, but had lost some of the asymmetry of the medial phalanges. From this point on, canids never regained the ability to retract their claws. Their claws protrude during all stages of walking and running, so the claw sheaths become blunt because of excessive wear on the ground. (One can often hear a dog's footsteps as its claw tips hit the ground, but not a cat's footsteps, which are silent because the claws are retracted).

Blunt claws are probably not a great hindrance when dealing with prey that are substantially smaller than the predators themselves. These prey can be overpowered by sheer size. When it comes to grappling with larger prey, however, particularly those that are much larger than the predators themselves, the lack of sharp claws can be a shortcoming. In this regard, canids seem handicapped when compared with felids, which as individuals are more capable of securing prey. Such a handicap may be one of the reasons why canids are primarily social hunters, which gives them an advantage over prey (chapter 5).

Neck

In recent years, research by Mauricio Antón (Antón and Galobart 1999) has demonstrated the importance of neck bones and musculature in mammalian predators. There has been considerable variation in the relative length and muscularity of the neck among both extant and extinct dog species, and some of the observed differences are likely to reflect different adaptations. Most species of modern canids have relatively long, straight necks, well muscled but relatively weaker than cats' necks, which are short and have a strong S curve. Due to the properties of muscle tissue, a shorter muscle contracts more efficiently than a longer muscle, so, other things being equal, a shorter neck can exert more power with less effort. That is why cats' short necks contribute so well to their positioning the prey's head and keeping it in place during the killing bite. This function was further emphasized in saber-toothed cats due to the large size of their prey and the need to avoid unexpected stresses during the bite, which could damage the cats' fragile upper canines.

Not only are dogs' necks longer than cats' necks, but the processes for muscle insertion on each cervical (neck) vertebra are smaller in dogs, indicating less powerful muscles. One way in which modern dogs compensate for the mechanical disadvantage of their long necks is the possession of the *nuchal ligament*, comparable to the one found in ungulates (figure 4.17). The nuchal ligament stretches along the back of the neck from the tips of the anterior thoracic vertebrae, and it helps to support the weight of the head without the need for active muscle contraction, thus saving energy. (There is one important difference, though: in ungulates, the ligament attaches on the back of the skull, but in dogs it attaches on the back of the axis, or second cervical vertebra, so the term *nuchal* is not so appropriate for this feature in dogs.) No other extant carnivore has a nuchal ligament, and it is intriguing to speculate about the origin of this structure in dogs. Fossils show that some dogs had relatively shorter necks, however, with more developed muscle insertions, and, judging from the morphology of their axis vertebra, lacked the nuchal ligament.

Early members of the subfamily Caninae were the first to show clearly the modern neck morphology associated with the presence of the nuchal ligament, suggesting that the structure is unique to this canid subfamily. Another distinctive trait of such early canines as *Eucyon* was the lengthening of the leg bones, especially the distal elements of the forelimb, such as the radius and metacarpals. As we have seen, this adaptation occurred for more efficient running, and in the case of *Eucyon* it most likely had to do with the need to forage across larger territories in search of prey because *Eucyon*'s habitat was becoming drier and more open, and prey were ever more dispersed. It is tempting to see a connection between the development of longer legs and of a longer neck, but other carnivorans with long legs, such as the cheetah, have short necks. Can we still see a functional connection between these changes?

Indeed, dogs' foraging method differs from cats' method in being more scent oriented. Cats depend largely on their eyesight and hearing to detect prey, but, as we know, dogs must "follow their noses" to locate prey. As early canines' legs became ever longer, a longer neck was necessary to allow them to follow scent trails with their nose close to the ground. The nuchal ligament became an advantageous feature in allowing canines to keep this posture for long periods of time, and the neck muscles in general became reduced as a way to save weight and energy. Reduced muscle power in the neck was not a great problem for these early canines, which hunted relatively small prey. Larger, later members of the family paid a price for the loss of strength in their neck muscles and for the feebleness of their forelimbs, which, although better suited for running, were less suited for handling prey. Nonetheless, these shortcomings were amply compensated for by group-hunting techniques, which all the larger, hypercarnivorous species of modern dogs share.

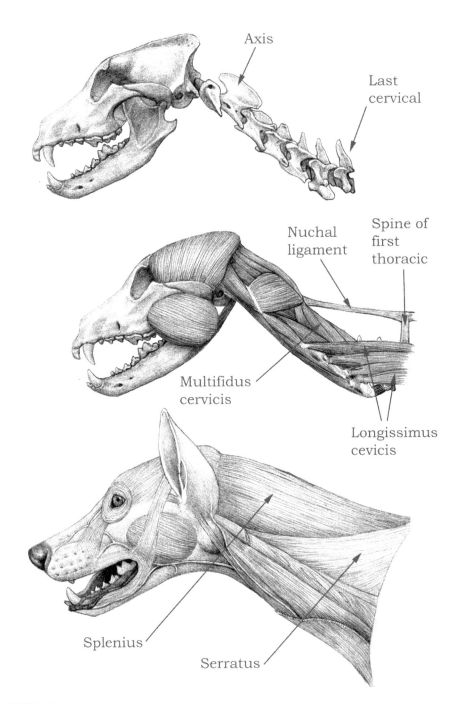

Axis

Last
cervical

Nuchal
ligament

Spine of
first
thoracic

Multifidus
cervicis

Longissimus
cevicis

Splenius

Serratus

FIGURE 4.17

Gray wolf

Skull and cervical vertebrae (*top*), deep muscles (*center*), and superficial muscles of the neck and head (*bottom*) of the extant gray wolf (*Canis lupus*). The neck is long and straight, and the nuchal ligament (extending between the spines of the anterior thoracic vertebrae and the back of the axis, or second cervical) helps to keep the neck raised without much muscular effort.

A different condition has been found in the fossils of some species of hypercarnivorous dogs. The neck of *Aelurodon* is a dramatic example (figure 4.18). Compared with a modern wolf (see figure 4.17), *Aelurodon* displays a remarkably short, S-shaped neck, with developed processes for muscle insertion, and the shape of its axis suggests the absence of the nuchal ligament. Such features would have been advantageous for such a hunter of large prey and, coupled with relatively shorter, more flexible, and more muscular forelimbs, would have made it more capable of dealing with large prey when hunting either by itself or in small groups. Such a design was easier to develop because the ancestral borophagines had retained a relatively primitive neck design, lacking the nuchal ligament. Modern wolves, dholes, and African hunting dogs would also benefit from such a cervical anatomy, but they are constrained by the morphology of their smaller ancestors. Even so, a detailed study of skeletal proportions (Hildebrand 1952) showed that these three species of big-game hunters have slightly shorter necks than would be expected for their size, when compared with their smaller relatives, such as jackals and coyotes. It would seem, therefore, that some "evolutionary adjustment" is under way.

Brain

Although brains are almost never preserved in the fossil records, they leave a distinct impression on the inner surface of the braincase in the form of ridges and grooves (*gyri* and *sulci*, in specialist terminology). Paleontologists can make latex endocasts (replicas) of these impressions, and the results are fairly detailed brain models for comparative purposes. Although paleontologists described natural molds of *Hesperocyon* 70 years ago (Scott and Jepsen 1936) (when braincase bones are broken in just the right way, the brain morphology is often revealed by such a mold), Leonard Radinsky (1969, 1973) was the first to explore systematically the comparative brain morphology in canids.

The ancestral canid *Hesperocyon* from the late Eocene to the early Oligocene (35 to 30 Ma) had relatively simple foldings and an underdeveloped frontal region of the brain. This was probably a primitive condition for all carnivorans. By the early Miocene (22 Ma), most canids had an expanded prefrontal cortex of the orbital gyrus immediately behind the olfactory bulb, and this trend continued with further expansions in later canids. Radinsky noticed the presence of a dimple in the coronal gyrus (near the frontal region) in three hypercarnivorous species of social hunters: *Canis lupus*, *Cuon alpinus*, and *Lycaon pictus*. He speculated that a relatively large prorean gyrus in the prefrontal cortex may be related to pack-hunting behavior. Although such studies may suggest interesting patterns, we caution that behavioral patterns are often difficult to correlate with external brain morphology.

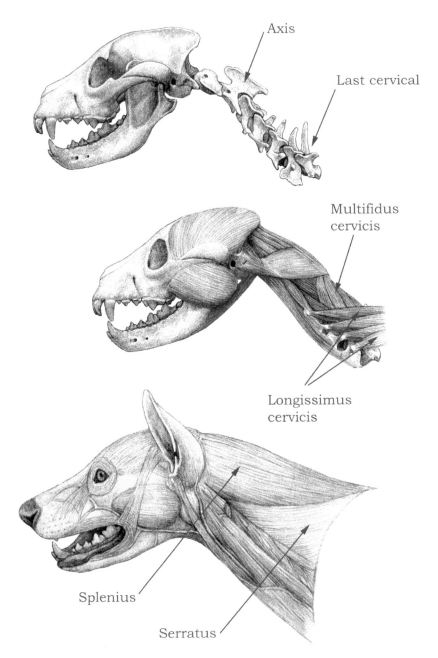

Axis

Last cervical

Multifidus
cervicis

Longissimus
cervicis

Splenius

Serratus

FIGURE 4.18

Aelurodon mcgrewi

Skull, cervical vertebrae, and neck muscles of the borophagine dog *Aelurodon mcgrewi*. Not only were *A. mcgrewi*'s neck muscles stronger than the wolf's, as shown by the larger processes for attachment in the vertebrae, but the smaller length and the strong S curve of the neck in *Aelurodon* imply that deep extensor muscles, such as the multifidus cervicis, had a much shorter span from origin to insertion and thus offered a great mechanical advantage. Processes for insertion of other muscles that extend the neck and flex it to the side, such as the longissimus cervicis, extended farther to the sides in *Aelurodon*, allowing these muscles to twist the neck more efficiently. Even the more superficial extensors, such as the splenius and serratus muscles, were more efficient thanks to their shorter spans.

John L. Gittleman (1986) did an extensive study of relative brain sizes of various carnivorous species and found a significant difference among families of Carnivora. Canids are relatively "brainier" than most other carnivorans except for the ursids. Gittleman attributed such a difference to historical developments in various families of the Carnivora during their early evolution. He also found that the more carnivorous groups (those that ate more meat than other foods) also had larger brains. He speculated that increased brain size in carnivorous species was due to their more complex foraging strategy, involving selection for rapid prey detection, pursuit, capture, and consumption.

5 HUNTING AND SOCIAL ACTIVITY

AS PREDATORS, CANIDS MUST HUNT PREY on a regular basis, just as any other group of carnivorans that consume meat as a significant component of their diet. Hunting is of paramount importance in any predator's daily survival. It is perhaps not surprising that the way certain species of carnivorans hunt largely defines their ecologic niche. Canids' hunting techniques and the issues of social hunting have come to define canids as a group.

Why Size Matters

Several important issues are related to body size among carnivorans. As anyone who has a cursory experience with contact sports knows, size matters. Sheer size can overpower an opponent no matter how skillful that opponent is. This is why sports such as wrestling and boxing are divided into weight classes—a significantly heavier opponent has too much of an advantage. When it comes to push and shove, the universal law of physics is the biggest equalizer in the world of predators and prey.

In the animal world, size is often a matter of survival. Giants such as adult elephants and rhinos are essentially free from worries about predators (their babies are more prone to predation). This is possibly the most important reason for the repeated developments of gigantism in the history of vertebrate evolution. Simply put, giant herbivores pose too much of a risk for smaller carnivorans: in choosing who is to become the next meal, carnivorans must constantly balance the benefit of a large, sumptuous meal with the potential hazard of injuries associated with tackling someone larger than themselves. Therefore, the interplay of body sizes between predators and prey is one of the most fascinating fields of inquiry in our understanding of predator–prey dynamics.

Although sheer size in herbivores certainly offers an undeniable benefit in terms of predator avoidance, considerations for the predators can be significantly different. In 1999, a group of British zoologists led by Chris Carbone marshaled empirical evidence that prey choice is often determined by the *predator*'s body size. By recording the body masses of predators and their prey, Carbone's group revealed a striking difference between small predators of less than approximately 21 kg and large predators of greater than 21 kg. Small predators largely prey on invertebrates or small vertebrates whose body size is substantially smaller than their own, whereas large predators often prey on vertebrates of equal size or larger. There appears to be a body-size threshold of around 21 kg, above which a predator must begin to tackle prey larger than itself. Why is there such a demarcation? Why don't large predators prey on small invertebrates and vertebrates in order to avoid injuring themselves?

The physical difficulties for small predators in handling large prey may be a significant reason for the dichotomy in what different size predators hunt and eat, but Carbone and his colleagues (1999) argue that energy is the key consideration. They suggest that small predators can sustain themselves on small invertebrates because of their low absolute energy requirements. In contrast, larger carnivores must consume a greater quantity of energy-rich food to support their daily activities. These researchers hypothesize that this body-size dichotomy is the consequence of body mass–related energy requirements; a body size of 21 kg is the critical point where a predator must switch from a diet of invertebrates and small vertebrates to a diet of larger prey. Large predators burn a large number of calories in pursuit of prey, a loss of energy that must be balanced by a greater intake of food. Thus as carnivorans become larger, they must make a change in their predatory habit in order to acquire enough calories for their daily energy expenditures. This hypothesis may sound suspiciously like a reversal of the popular cliché "You are what you eat," but in this case the saying should be "What you eat actually depends on what you are." The picture becomes even more complicated from an evolutionary perspective.

The Games of Becoming Bigger

Pioneer vertebrate paleontologist Edward Drinker Cope (1880) articulated a "law" that states that vertebrate evolution tends to proceed from small to large body sizes. Two extreme cases of "Cope's law" are the dinosaurs in the Mesozoic (248 to 65 Ma) and the whales in the modern seas. Although biological evolution almost never operates in a lawlike fashion in the sense that every organism must obey this law (exceptions abound in body-size reductions in many lineages), Cope's law can be viewed as a kind of rule of thumb in many lineages and thus might be best called "Cope's rule."

In light of predator–prey dynamics, Cope's rule does seem to apply in many lineages. Blaire Van Valkenburgh, Xiaoming Wang, and John Damuth (2004) have explored Cope's rule as it applies to canid evolution, demonstrating that canids have indeed repeatedly evolved larger and larger species in many lineages and, intriguingly, that the larger-bodied species also tend to become more hypercarnivorous in their jaws and dentitions (chapter 4). In other words, there appears to be a pervasive selection for larger size in canids, and as they become bigger, their diets also become more restricted to meat. Although we concentrated on published data on the hesperocyonine and borophagine canids (figure 5.1), we suspected that the same conclusion also applies to the canines, particularly in the *Canis–Xenocyon–Cuon–Lycaon* lineage. Our evolutionary perspective on the fossil history of canids is thus consistent with the body-size constraints hypothesized by our zoologist colleagues.

We hasten to add that not *every* canid lineage is subject to such an evolutionary selection. In fact, throughout canid evolution there have been persistent small, fox-size lineages that remained small and from time to time even reduced their body size. These foxlike canids tend to be evolutionarily conservative not only by staying small, but also by changing little in other parts of their bodies. It is possible that this conservatism among small canids was the result of competition from larger species; established lineages of larger predators may have prevented others from entering the niche of large predators. Small canids may not have occupied flashy positions in the pecking order (in the sense of interspecific competitions, not intraspecific competitions), but they have been important in the overall health of a particular family of carnivorans. When a lineage of large predators became extinct, small and opportunistic species almost invariably quickly filled the niche by becoming larger as competition was relaxed momentarily. Such was the case for such foxlike forms as *Hesperocyon*, *Archaeocyon*, and *Leptocyon*, which were able to capitalize on the extinction of other hypercarnivores. The peril of lacking small species in a lineage is seen in the hesperocyonines and borophagines. Large, fierce predators such as *Osbornodon* and *Borophagus* may have been without competitive peers, but when both became extinct, no small hesperocyonines or borophagines were around to continue the lineage, and both subfamilies became extinct along with their top predators (see figure 7.2).

Along this line of thought, it may be interesting to pause for a moment to consider the future of the current crop of carnivorans. With the exception of the Hyaenidae, all other families in the order Carnivora have a balanced spectrum of small and large species. If the large species were to become extinct, smaller forms would probably quickly evolve to a larger size to replace the extinct forms. Hyaenids are the only group for which this probably would not be true. With the exception of the aardwolf (*Proteles cristatus*), which is a highly specialized termite eater, all three

FIGURE 5.1

Comparison of *Epicyon haydeni* and *Archaeocyon leptodus*

Skulls and reconstructed heads of two borophagine dogs, drawn to the same scale. *Top left* and *center*, *Epicyon haydeni*; *top right* and *bottom*, *Archaeocyon leptodus*. *Archaeocyon*, from the late Oligocene (27 Ma), is the earliest and most primitive member known of the subfamily, and *Epicyon*, from the late Miocene (9 Ma), included the largest borophagines and, in fact, the largest dogs ever known.

living species of hyenas are large hypercarnivores. If they become extinct, the great evolutionary lineage of hyenas will end without a smaller form within the Hyaenidae to replace them.

Sexual Dimorphism

The term *sexual dimorphism* refers to morphological differences between the sexes of a species. For example, human males in the United States average 78.5 kg, whereas human females average 62 kg. This difference translates to a 26.6 percent dimorphism in male/female body mass, which is modest compared with the dimorphism of other species, such as the notoriously dimorphic elephant seals (*Mirounga*), whose males can be two or even three times heavier than the females.

When it comes to canids, sexual dimorphism is often quite modest. For example, measurements from our own data yield the following results. The average skull length of a population of 12 New England red foxes (*Vulpes vulpes*) was 114.4 mm for the males compared with 106.4 mm for the females; that is, males are 7.5 percent larger than females in linear measurements. The average skull length of a population of 12 Nevada coyotes (*Canis latrans*) was 145.2 mm for the males and 142.9 mm for the females, yielding an even smaller dimorphism of 1.6 percent. In general, however, larger, more hypercarnivorous species tend to be slightly more dimorphic than smaller, hypocarnivorous forms. Studies published by various authors suggest a 3 to 6 percent linear dimorphism in cranial and dental measurements in *Vulpes*, 3 to 8 percent in *Canis*, and close to no dimorphism in *Urocyon* (gray fox) (Gingerich and Winkler 1979; Gittleman and Van Valkenburgh 1997).

For extinct taxa, distinguishing male and female canids in fossils is generally difficult, except in the rare instances in which a baculum (penis bone) is preserved within a skeleton, which is a sure sign of a male. The most dimorphic part is commonly the size of the canine teeth. Males tend to have relatively larger canines than females (which may be due to competition for mates; males that display larger canines when snarling may have a better chance of scaring off the competition, as is common in many male vertebrates that sport conspicuous display appendages, such as horns, antlers, and tusks). We thus use canine teeth size as an imprecise criterion for judging sex in fossil forms. Based on this method for sorting fossil materials, it is speculated that extinct canids had a small range of sexual dimorphism, similar to that of their extant counterparts.

In mammals, highly dimorphic species tend to have a polygynous mating system, in which a dominant male mates with multiple females in a pride, to the exclusion of other males. Modern African lions and elephant seals typify such a system. In contrast, canids are more often monogamous in social groups and, as a consequence, do not need a large dimorphism to sustain such a system. This was true even for some of the largest canids of the past. Based on measurements of the extraordinary collection of dire wolves (*Canis dirus*) at the George C. Page Museum of Los Angeles, Blaire Van Valkenburgh and Tyson Sacco (2002) estimated that the giant Ice

Age dire wolf had a level of sexual dimorphism comparable to that of large extant canids.

Social Hunting

Although the members of many fox species are solitary, most canids, particularly those of large body size, have developed highly complex social behavior. Hypercarnivorous species such as the African hunting dog and gray wolf form complex packs organized mostly by family groups of up to 30 individuals. In a pack, the dominant pair breeds, with the less dominant females under behaviorally induced reproductive suppression. Adults of both sexes also contribute to parental care in guarding and feeding the pups, a phenomenon known as alloparenting. During hunting, the members of canid packs cooperate in relays, ambushes, and other teamwork. With the possible exception of hyaenids, such a complex social structure is unrivaled by most other families of carnivorans.

It is therefore relevant to ask: Why is it advantageous to hunt in a group? After all, living in groups does have its costs. Parasites are more easily spread in a group (thus often requiring mutual grooming), the spoils of group hunting must be shared, and the effort of raising the young must also be shared. Group hunting does have advantages, however, and behavioral ecologists have defined three factors in its favor—strength in numbers, protection of kills, and defense of territory—to which we can add functional morphology.

STRENGTH IN NUMBERS

The advantage of ganging up on any opponent is fairly obvious. Pack hunting offers higher potential success through collective efforts such as relays and ambushes (figure 5.2). Perhaps more important, pack hunting allows a group of canids to tackle prey that is many times the size of any individual predator. The sheer size of a large artiodactyl—such as a musk ox, bison, or moose—is normally beyond the strength of a single wolf. In a pack, however, wolves are emboldened to tackle just about any large mammal they encounter. Even large carnivorans, such as bears, that individually may reign as kings of their territories have been known to be chased by pack wolves. Pack hunting in wolves is often a long-distance chase that includes wearing down the prey and injuring it with small bites, a prolonged affair that sometimes takes hours. Such a long-distance chase by a solitary hunter is difficult to imagine. Even when hunting small prey, although a single predator is enough to overwhelm the prey, collaboration can still be beneficial by employing such tactics as encircling the prey and cutting off its retreat.

FIGURE 5.2

A scene in Spain during the early Pleistocene

A pack of canids of the species *Xenocyon lycaonoides* is in pursuit of a goatlike antelope. The great similarity between this extinct species and the African wild dog (*Lycaon*) leads us to infer a similar hunting style.

PROTECTION OF KILLS

In a balanced carnivoran community such as in Africa, raiding someone else's prey can be an easier way of obtaining a meal than trying to capture the prey by one's own effort. Even the largest canid in Africa, the hunting dog, which weighs 20 to 30 kg, is individually no match for a lion or a spotted hyena. Consequently, these other species often loot hunting dogs' food. The lost food results in significant increases in the hunting dogs' daily energy expenditures because they have to make up for the loss, a factor that may have contributed to the relative rarity of hunting dogs in Africa. A pack, however, not only offers protection against other species, but also guards against other packs of its own species.

DEFENSE OF TERRITORY

Large, pack-hunting canids are generally highly territorial. To defend the resources within the pack's own territory increases the economy of a pack. Gray wolf packs can have home ranges of up to 6,000 km², although the range varies greatly. In such a large territory, chance encounters with members of other packs or lone wolves are rare, but when they do happen, resident packs defend their territories fiercely; strangers are often killed without mercy.

FUNCTIONAL MORPHOLOGY

Beside these three factors, functional morphology also offers a compelling case for social hunting. As discussed in chapter 4, canids lost the ability to retract their claws early in their history. Compared with the retractile claws of felids, the relatively blunt claws that dangle from the tips of canids' digits are relatively ineffective in grasping and hanging onto prey. Rather than making a quick kill by jumping on the prey and delivering a killing bite, as felids do, canids may have to resort to relatively ineffective bites on the rump and thereby wear down the prey more slowly. Therefore, when it comes to preying on animals much larger than themselves, canids have to rely on their social group.

Whatever other advantages (or lack of advantages) there may be in the formation of packs, the fact that some large living species (for example, gray wolf, African hunting dog, and Asian dhole) consistently form packs implies that canid sociality has offered on balance significant benefits for the family Canidae as a whole. Given this fact, one is also tempted to ask, as a corollary: Does increased social organization also mean a corresponding increase of intelligence? In other words, is there any relation between social hunting and intelligence? Are complex social behaviors passed down from generation to generation? Are canids more intelligent than other carnivorans? More specifically, is the dog more intelligent than the cat?

If these questions seem difficult to answer for living species (chapter 8), they are much more difficult to address for extinct taxa. The most obvious place to look for the answers is the brain. Brains, however, do not fossilize, and paleontologists are forced to look at the endocasts as a substitute. The vertebrate paleontologist Leonard Radinsky (1973) may have been the first to attempt to answer the question of social hunting among extinct species of canids by studying the endocasts of fossil canids from various periods. Radinsky observed that canids' brains have a larger prorean gyrus than felids' brains. The prorean gyrus is a bulge located in the front part of the brain. Modern pack-hunting canids—such as *Canis*, *Cuon*, and *Lycaon*—seem to have an even more dorsally expanded prorean gyrus than do foxes. When Radinsky examined the fossils of large borophagines—such as *Aelurodon*, *Epicyon*, and *Borophagus*—he found that these structures were not well developed, although he stopped short of speculating whether these large canids were pack hunters or not.

Blaire Van Valkenburgh, Tyson Sacco, and Xiaoming Wang (2003) addressed the question of social hunting in a different way. Among living species, hunters of large prey are often associated with a suite of adaptations, including deep jaws, broad muzzles, enlarged incisors, and enlarged canines. These features seem to be adaptations for dealing with the heavy loads placed on their skulls and teeth when they are hunting and killing large prey. By quantifying these morphological features in a morphometric analysis using the statistical methods of principal component analysis, we found that large borophagine canids occupied an intermediate morphospace between living hyenas and canids. Like hyaenids, hypercarnivorous borophagines had stronger jaws and more enhanced jaw muscle leverage than do living canines. Unlike hyaenids, but like canines, borophagines retained a substantial grinding dentition in the postcarnassial molars. It seems clear that although saddled with the phylogenetic constraints of being a canid, the large, hypercarnivorous borophagines resembled pack-hunting hyenas and African wild dogs. Given that borophagines have no retractile claws, it is reasonable to assume that they may have hunted in packs (figure 5.3).

FIGURE 5.3
Aelurodon ferox

Two borophagine dogs of the species *Aelurodon ferox* are in hot pursuit of their prey. The dentition, anatomy, and body size of such large borophagines suggest that they were hunters of large prey and that they may have joined in groups of two or more to pursue large ungulates.

Scavengers or Hunters?

Many borophagines were earlier regarded as hyaenoid dogs, based on their numerous cranial and dental similarities to modern hyaenids; these similarities implied that carcasses, rather than fresh kills, were the main diet of these extinct canids. In popular sentiment, scavengers often have an unsavory reputation as chasers of rotten meat and leftover scraps. They are also relegated to a less important ecological role in evolution because they do not directly affect prey communities, in contrast to hunters, which directly influence the evolution of their prey by forming a predator–prey relationship. A closer look at the strongest bone cracker in the modern East African community, the spotted hyena (*Crocuta crocuta*), reveals, however, that the species hunts for fresh kills far more frequently than it scavenges on carcasses. The strong premolars in spotted hyenas help them quickly consume their prey in group feeding frenzies, with the bones and flesh gone in a matter of minutes. Smaller hyenas, such as the brown hyena (*Parahyaena brunnea*) and striped hyena (*Hyaena hyaena*), in contrast, do consume carrion as a more significant portion of their diet.

The question of scavengers versus hunters is even more difficult to answer with regard to extinct carnivorans, a problem that also plagues dinosaur paleontologists in their controversies over the diet of *Tyrannosaurus rex*. Did large, bone-cracking forms such as *Enhydrocyon*, *Epicyon*, and *Borophagus* vigorously pursue prey of their own or search for carrion? Kathleen Munthe (1989) concluded in a doctoral study of functional morphology of borophagines that the robust skulls, flexible forelimbs, and often heavily worn teeth of large borophagines such as *Borophagus secundus* were adaptations for scavenging. However, strong skulls and blunted teeth are as much an indication of bone cracking as they are characters of scavenging, a connection that is often inadequately appreciated. A strong set of teeth capable of cracking large bones certainly helps to make use of all nutritional value of carrion, but, as evidenced in living hyenas, not all bone crackers are scavengers. Perhaps a high degree of bone-cracking specialization is more an indication of social feeding, where speed of consumption is a premium. If so, fossil forms with bone-cracking teeth may be more indicative of a high degree of social hunting.

Scavengers also generally have a low density and a restricted distribution because of the relative scarcity of resources (carrion), as exemplified by some modern scavenging carnivorans, such as the brown and striped hyenas. Fossil records for the borophagines indicate, however, the opposite of what may be expected for scavengers. Some advanced borophagines, such as *Borophagus secundus*, were so abundant in certain quarries (for example, in the late Miocene Coffee Ranch Quarry in northern Texas) and so widespread in North America (as far south as Honduras and El Salvador) that there is little doubt of its dominant status in the carnivoran fauna of those

FIGURE 5.4

Borophagus secundus

A pack of dogs of the species *Borophagus secundus* are feeding on their prey. Competitive group feeding as depicted in this reconstruction would have favored complete utilization of the carcasses, with individual dogs taking pieces of the carcass away from the group for more quiet consumption and cracking the bones to reach the nutrients within.

areas. It is difficult to imagine that carrion feeding alone could have sustained such a large population (figure 5.4).

Canids Versus Felids

Three modern families of carnivorans include top terrestrial predators: the Canidae, the Felidae, and the Hyaenidae (the list excludes the Ursidae because although bears can be dominating in certain situations, they are largely omnivores and, with the exception of the polar bear, rarely use meat as the main part of their diet year round). The top predators in each group are often large hypercarnivores that are capable of hunting down prey several times their own size. Although not necessarily the all-time champions in their own lineages (each family has extinct members in distant history—such as *Epicyon* in the canids, *Pachycrocuta* in the hyaenids, and various saber-toothed cats in the felids—that were larger, more powerful, and more specialized than their modern counterparts), the current crop of elite predators represents the ultimate killing machines among their living relatives. It is thus instructive to compare the hunting behaviors among these three groups of great predators. Through such a comparison, one can gain additional appreciation of the ways in which large carnivorans behave.

At the risk of oversimplification, the comparison of canids and felids can be summarized as a contrast between endurance and speed in canids, on the one hand, and stealth and power in felids, on the other (for more on felids, see Turner and Antón 1997). Cats are often referred to as the ultimate killing machines because of their extraordinary combination of speed, agility, grace, and killing power. Such a formidable combination of anatomical traits is ideally suited for animals with hypercarnivorous dentitions that eat meat exclusively. The ability to climb trees (arboreal locomotion) is universally present in felids (at least among those species not too large to climb), perhaps because of their preference for wooded environments. Trees offer excellent camouflage, but they can also be dangerous obstacles during a fast chase. It thus makes intuitive sense that felids almost invariably approach their prey in stealth and try to pounce on it in surprise attacks. Capture is often accomplished by securing the prey with sharp claws and delivering a lethal bite with powerful canines (figure 5.5). The chase is short, and the kill is quick.

Modern canids, by contrast, seem to have descended from a terrestrial ancestor adapted to open plains (some living species may have adapted to wooded environments, such as the North American gray fox [*Urocyon*]) (chapter 6). Canid limbs are long, too, but more emphasis is placed on long-distance endurance than on a short burst of speed. A surprise attack is seldom achieved in an open plain. Therefore, it is less important to subdue the prey in the shortest possible time than to outrun and exhaust the opponent. Lacking retractile claws, a powerful weapon for most felids, canids rely more on social hunting when confronting large prey (although lions are also capable of complex cooperative hunting). Such a combination of anatomical and behavioral characteristics, including versatile, mesocarnivorous dentitions, permits canids to be highly opportunistic in their diet.

These comparisons are, of course, generalizations about two families that have diverse members. However, family membership appears to place strong constraints on behavioral repertoires. These constraints are phylogenetic, in that close genealogical relationship tends to be the best predictor of behavior patterns. About phylogenetic constraints, behavior ecologists David W. Macdonald and Claudio Sillero-Zubiri comment: "[H]aving dwelt on the burgeoning discoveries of diversity amongst

FIGURE 5.5
Lion

A modern lion hunts an antelope in two stages. Large cats are well adapted to deal individually with large prey by holding and bringing down their victim with their strong, clawed forepaws and then dispatching it with a powerful killing bite. Canids are not anatomically suited for such a hunting style.

the canids . . . our lifetimes spent watching these creatures have simultaneously and paradoxically led to a realization of their sameness" (2004:6). In many ways, the contrast of canids and felids in this chapter parallels a stereotypical "dog versus cat" comparison. After all, both domestic animals are recent descendants of wild species (chapter 8), and their behavior closely mirrors that of their wild ancestors.

Canids Versus Hyaenids

Hyaenids seem to be much more similar to canids than to felids, both behaviorally and anatomically. A casual observer cannot help seeing the overall similarity in their appearance and is thus often surprised to learn that hyenas are actually more closely related phylogenetically to cats than to dogs. Indeed, all modern hyenas are "dog-like" in size, body proportion (except the uniquely stooped shoulder), and length of rostrum. A closer examination reveals additional similarities in both morphology and behavior. Like their canid counterparts, hyenas have blunt, nonretractile claws and therefore do not climb trees. As a result, their teeth are the only tool they can use to secure their prey, and claws are never part of their weaponry. Like canids, hyaenids have a full complement of premolars arranged in a long rostrum and jaw, in contrast to the reduced premolars and short rostrum of felids. Hyenas kill their prey by consuming them alive rather than by delivering a killing bite on the neck, as do felids. Also like the canids, hyaenids are persistent pursuers rather than stalkers in ambushes and are great long-distance chasers. Given such a combination of capabilities and limitations, it is not surprising that hyenas also tend to be highly social hunters in packs of up to 25 individuals, particularly when trying to overcome larger prey.

The similarity between canids and hyaenids is a good example of convergent evolution over a long geological history. Hyaenids and canids embarked on their separate evolutionary pathways almost right after the beginning of the Cenozoic (about 65 Ma), when dinosaurs had just become extinct after a catastrophic impact by an asteroid. Shortly after the beginning of the Cenozoic, there appeared the early progenitors of viverravids, which are distally ancestral to all feliform carnivorans, including the Hyaenidae. (A study by Gina D. Wesley-Hunt and John J. Flynn [2005] suggests, however, that viverravids, as an early offshoot of carnivorans, did not give rise to any living family.) Miacids, the progenitors of caniform carnivorans including the Canidae, appeared in the Eocene a few million years later.

The earliest members of the family Canidae, *Prohesperocyon* (chapter 3), appeared in the late Eocene (around 40 Ma), becoming the first family of modern carnivorans to emerge in North America, as indicated by the fossil record. The Hyaenidae, however, appeared much later, in the early Miocene (around 20 Ma) of Eurasia. Hyaenids thus have half as long a history as canids. Although hyaenids occupied a much larger and diverse theater in Eurasia than canids did in North America, the overall diversity of hyaenids through time was substantially lower than that of canids. Nonetheless, at their peak in the late Miocene and the Pliocene (8 to 3 Ma), hyaenids included an array of predator species from giant bone crackers to swift pursuit predators similar to such early canids as the hesperocyonines and borophagines. Canids and hyaenids are ecologically probably closest to each other among both extinct and extant forms, and they have independently evolved a number of similar features. Therefore, when it comes to finding an analogue for such large borophagines as *Aelurodon* and *Epicyon*, living spotted hyenas probably offer the best example for how borophagines used their massive premolars to crack bones.

PLATE 1

A scene in North America during the late Eocene

An adult *Hesperocyon gregarius* watches over her two pups in front of their den, in a forest environment.

Life reconstruction of *Enhydrocyon*

A large hesperocyonine of the genus *Enhydrocyon*, from the late Oligocene to the early Miocene (Arikareean) of North America, uses its strong premolars to crush the bone of an ungulate.

PLATE 3

Life reconstruction of
Osbornodon fricki

Osbornodon fricki, from the early
Miocene (Hemingfordian) of North
America, was the largest member
of the subfamily Hesperocyoninae,
reaching the size of a small wolf.

PLATE 4 (FACING PAGE)

A scene in western North America during the Miocene (Barstovian)

A pack of the wolf-size borophagine *Aelurodon ferox* is in pursuit of a three-toed horse of the genus *Neohipparion*.

PLATE 5

Life reconstruction of *Borophagus secundus*

The large *Borophagus secundus*, from the late Miocene (Hemphillian) of North America, was an advanced borophagine that showed the most extreme adaptations for bone crushing in its strong skull and robust premolars.

PLATE 6

A scene in North America during the late Miocene (Hemphillian)

In an environment of grassland and forest patches, a young adult *Eucyon davisi* approaches one of its parents in a submissive attitude. It is possible that such youngsters remained in their parents' territory and helped raise pups, which is the origin of larger social groupings in several species of the subfamily Caninae. Two antilocaprine antelopes of the genus *Texoceros* watch the jackal-size canids from behind the fallen tree.

PLATE 7

A scene in North America during the Pliocene (Blancan)

In this mosaic of grassland and woodland, a solitary adult *Borophagus diversidens* tries to defend its prey, the camelid *Hemiauchenia*, from a pack of *Canis lepophagus*. Facing only one member of this last species of borophagine would be the only chance for the smaller canids to appropriate a carcass. If *B. diversidens* had the help of other members of its species, the coyote-like *C. lepophagus* could only wait until the larger dogs had gorged themselves to satisfaction.

A scene in western North America during the late Pleistocene (Rancholabrean)

An adult dire wolf (*Canis dirus*) calls for its pack while a herd of Columbian mammoths (*Mammuthus columbi*) ambles past in the background.

6 CHANGING ENVIRONMENTS AND CANID EVOLUTION

AS MEMBERS OF THE ORDER CARNIVORA, the majority of canids are predators. Thus throughout their evolutionary history, canids have been closely inter-
twined with their prey, which in turn have been directly affected by the surrounding
plant communities.

During the past 20 years, a global picture of the long-term paleoclimatic record
has emerged. Through studies of air trapped in ice cores in polar regions, of drill
cores from oceans and lakes, and of wind-blown sediments, we can learn much
about the climatic histories of various regions. In particular, microorganisms such
as the single-celled foraminifera preserved in marine sediments permit us to obtain a fairly detailed picture of ocean temperatures in the past, which serve as an
approximate indicator of global environments (figure 6.1). From such studies, we
know that throughout the Cenozoic (most of the past 65 million years), numerous
fluctuations in temperature have profoundly affected these environments. The ab-
breviated narrative in this chapter attempts to give readers a sense of overall climatic
changes during the Cenozoic and its effect on the associated biological communities. Against this background, we can begin to appreciate the dramatic changes in
physical environments and the associated biological responses. Because canids were
confined largely to North America during much of their early existence, we need
concern ourselves with only their immediate environments in North America in
this era.

The Paleocene and the Beginning of Carnivora

The Cenozoic era immediately followed the extinction of the dinosaurs at the Cretaceous–Tertiary boundary (65 Ma) in a global catastrophe apparently caused by
the impact of an asteroid. After the dust settled, the beginning of the Cenozoic is

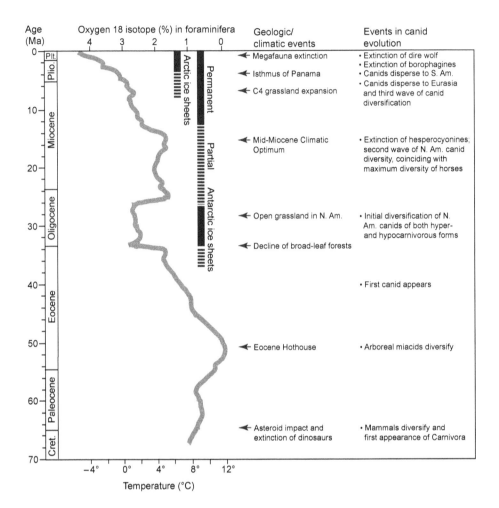

FIGURE 6.1

Events in the evolution of canids

Major events in canid evolution and corresponding geologic and biotic events, as associated with a global temperature curve (*scale on bottom*) during the past 65 million years throughout the Cenozoic (*vertical axis*). This paleotemperature curve (simplified from Zachos et al. [2001]) is compiled from oxygen isotopes (*scale on top*) preserved in bottom-dwelling, deep-sea foraminifera contained in sediments from multiple drill cores in various oceans. A clear trend of climatic deterioration from a global greenhouse to icehouse during the Cenozoic is readily apparent, and many of the more drastic temperature disturbances were triggered by major geological events. In an increasingly open landscape, canids thrived by becoming increasingly cursorial.

marked by a rather warm period in the Paleocene (65 to 55 Ma). During this epoch, mammals finally became the dominant land vertebrates, taking over many of the niches previously occupied by the dinosaurs. An explosive radiation of early mammals rapidly filled the ecological space left open by the dinosaurs' demise. Many of the Paleocene mammals increased in body size, as contrasted with their largely mouse- and rat-size ancestors in the Mesozoic (248 to 65 Ma), and some of them included the earliest members of lineages leading to modern forms, including the order Carnivora, which saw its first members in the early Paleocene in the form of the family Viverravidae (some interpretations of the fossil record suggest that carnivorans were present in the late Cretaceous [75 Ma] of North America; see "Cimolestids" [chapter 2]). However, the archaic viverravids featured a precociously reduced set of teeth and were probably not closely related to canids. During the late Paleocene (56 Ma), the ancestral group that gave rise to canids, the family Miacidae, also began to appear.

The Eocene Hothouse and the Origin of Canids

From the already warm and humid conditions in the Paleocene, the beginning of the Eocene (around 55 Ma) was marked by a rapid warming to a peak temperature more than 14°C higher than today's average global temperature. This extreme warming event is called the Paleocene–Eocene Thermal Maximum or, more colloquially, the Eocene Hothouse. With these warm conditions, the Eocene global climate was perhaps the most homogeneous within the Cenozoic as a whole. The temperature gradient—the differences between temperatures along the equator and temperatures at either pole—was only about half as much as it is today, resulting in a very equable climate with low seasonality. The climate was so warm that even the polar regions could support a diverse and productive biota, including the miacids.

The warm and humid conditions during the Eocene—coupled with a high level of the greenhouse gases carbon dioxide (CO_2) and methane (CH_4)—were ideally suited to the growth of dense forests in much of the world. During the Eocene Hothouse, tropical forest conditions expanded to the latitude of northern Wyoming, as recorded in the fossil plants from the Bighorn basin. Lush forest canopies dominated much of North America. It is perhaps no coincidence that primates, along with other forest-dwelling mammals, flourished under such conditions. Carnivores in the Eocene were similarly adapted to life in and around trees. A modestly diverse group of miacids began to explore various environments for opportunities, although they were mostly the size of small foxes or smaller and lived in the shadow of the hyaenodonts, a group of archaic predators that were not closely related to carnivorans and that were generally larger (coyote to wolf size) than the miacids and better equipped to prey on larger herbivores (chapter 2).

Through a series of small, foxlike miacids, the proto-canids gradually emerged. In the late Eocene (40 Ma), the first unambiguous canid, *Prohesperocyon*, appeared in southwestern Texas, represented by a single skull and lower jaw. *Prohesperocyon* probably did not differ substantially from other ancestral miacids in its way of life in the forest, but it bore the unmistakable marks of being a true canid: an inflated bony bulla covering the ear region and subtle dental features, such as the loss of the upper third molars. *Prohesperocyon*'s immediate descendant, *Hesperocyon*, appeared very soon thereafter in the northern Great Plains and Canada (figure 6.2). Fossil records of *Hesperocyon* are far more abundant than those of its *Prohesperocyon* predecessors. The profusion of fossil records of this early canid indicates its central role in the phylogeny of all canids. At this stage of canid evolution, it is difficult to point to a particular skeletal feature that may have given canids an overwhelming advantage, but a slightly more elongated limb with compressed digits seems to hint at their preadaptations to an increasingly open environment. However, their limbs were not fully digitigrade, and they were probably more suited for lives on the forest margin, where the ability to climb trees helped them escape formidable predators (figure 6.3). Nonetheless, the evolution of cursorial feet may have been a key adaptation that enabled canids to succeed in later, more open environments.

The Oligocene Cooling and the Initial Diversification of Canids

From the peak of the Paleocene–Eocene Thermal Maximum more than 50 Ma, the first major step toward a long trend of climatic deterioration during the Cenozoic occurred near the Eocene–Oligocene boundary (around 33.7 Ma), apparently associated with the initial appearance of ice sheets on the Antarctic continent. James P. Kennett (1977) proposed that plate tectonics was the ultimate cause of the formation of the Antarctic ice cap. He suggested that the formation of Antarctic circumpolar currents in the Southern Ocean thermally isolated the Antarctic continent from the world ocean. These circumpolar currents were initiated because of the initial tectonic opening of a plate boundary between Antarctica and Australia (Tasmanian Passage) and the later extension of this spreading center between Antarctica and

....

FIGURE 6.2

Comparison of size among mammals from the late Eocene (Chadronian)

From left to right: *Hesperocyon gregarius*, *Palaeolagus*, *Mesohippus*, and *Leptomeryx*. Few ungulate species attained large size during the Chadronian of the late Eocene (35 Ma), but not even the small species depicted here would be suitable prey for *Hesperocyon*, and only the early lagomorph *Palaeolagus*, along with rodents and other smaller vertebrates, would feature regularly in this early dog's menu.

FIGURE 6.3

A scene in western North America during the late Eocene (Chadronian)

The ability to climb onto thin branches would have helped the early canid *Hesperocyon* (*right foreground*), from 35 Ma, to escape from potentially dangerous, much larger animals, such as the saber-toothed nimravid *Hoplophoneus primeavus* (*left*) or the entelodontid *Archaeotherium mortoni*.

South America (Drake Passage). However, computer modeling of the climatic processes seems to indicate that a drop in atmospheric CO_2 may also be to blame for the formation of Antarctic ice (DeConto and Pollard 2003). Whatever the cause of the Antarctic ice sheet, global temperatures dropped by 4 to 5°C or even more within a short span of 1 to 2 million years. Such a significant temperature change caused a major faunal turnover among marine organisms, but the land animals were less affected (Berggren and Prothero 1992).

After the tropical forest conditions in the early Eocene, a steady decline in global temperatures throughout the Eocene gradually turned the plant communities into something close to warm-temperate forests. The sharp drop in temperatures at the Eocene–Oligocene boundary due to the isolation of Antarctica was associated with a shift from warm-temperate, angiosperm-dominated forest types to cooler-temperate, gymnosperm-dominated (mainly coniferous) forest types in high latitudes. In midlatitude North America, the climatic deterioration initiated a process of progressively drier conditions and more seasonal environments. Plant communities at midlatitudes responded to this trend of decreasing rainfall by changing from high-productivity, moist forests in the late Eocene to low-biomass, dry woodlands at the beginning of the Oligocene (34 Ma), progressing to wooded grasslands and ultimately to large areas of open grassland in the middle Oligocene (30 Ma) (Leopold, Liu, and Clay-Poole 1992).

The initial opening up of the continental interiors by the retreat of the woodlands and the development of a woodland-savanna-grassland mosaic landscape were the impetus for the evolution of the grassland vertebrate communities (figure 6.4; see figure 6.1). Mammalian herbivores began to develop high-crowned teeth and increased cursoriality. To a large extent, the history of canids is a story of the coevolution of a group of cursorial carnivorans with the emerging grassland communities. Carrying over from the late Eocene, *Hesperocyon* remained the only canid with a consistent fossil record during the early Oligocene. However, we also possess a few tantalizing fossils indicating that *Hesperocyon* had begun to diversify in the early Oligocene, seizing the opportunities afforded by the opening up of the landscape. This initial radiation of Canidae produced ancestral members of all three subfamilies of canids (Hesperocyoninae, Borophaginae, and Caninae). All these early progenitors were fox-size small canids, but each bore the morphological hints of later events. They suddenly appeared in the badlands of Nebraska, Colorado, South Dakota, and Wyoming during the Orellan age (34 Ma) at the beginning of the Oligocene. For the subfamily Hesperocyoninae, members of the *Mesocyon–Enhydrocyon* group and of the *Osbornodon* group emerged, and for the subfamily Borophaginae, the earliest member, *Otarocyon*, is recorded in the Orellan. For the subfamily Caninae, a single jaw fragment indicates its presence as a progenitor of *Leptocyon*.

FIGURE 6.4

A scene in western North America during the late Oligocene (Arikareean)

Left background, the large camelid *Oxydactylus wyomingensis*; *left foreground*, the canid *Enhydrocyon crassidens*; *right center*, the oreodontid *Leptauchenia decora*, from 27 Ma.

The rest of the Oligocene saw the continued diversification of canids and the first dominant dogs in the Hesperocyoninae's history, such as *Sunkahetanka*, *Philotrox*, *Enhydrocyon*, and *Paraenhydrocyon*. These hesperocyonines evolved highly hyper-

FIGURE 6.5
Comparison of size among mammals from the late Oligocene (Arikareean)

From left to right: the canids *Mesocyon coryphaeus, Archaeocyon leptodus, Enhydrocyon crassidens*, and *Phlaocyon leucosteus*; the protoceratid *Protoceras skinneri*; the oreodont *Leptauchenia decora*; and the equid *Miohippus gidleyi*. Horses such as *Miohippus* were among the largest potential prey for dogs in the Oligocene (27 Ma), and catching them would normally have required group action. Smaller ungulates such as the protoceratids and the abundant *Leptauchenia* could have been taken even by lone individuals of the largest dog species. Ungulate prey was essentially out of range for the smaller dog species, which would have taken rodents and other small mammals, as well as, in the case of *Phlaocyon*, a large proportion of nonvertebrate food. Reconstructed shoulder height of *Miohippus*: 70 cm.

carnivorous dentitions that are comparable to the dentition of the modern African hunting dog, and they achieved the body size of small wolves to begin hunting prey larger than themselves. In contrast, facing competition from the dominant hesperocyonines, the early borophagines tended to evolve toward hypocarnivorous dentitions and small body sizes to exploit less predaceous lifestyles. These borophagines were represented by *Archaeocyon, Cynarctoides, Phlaocyon*, and others (figure 6.5). Meanwhile, the canines were still biding their time with a few inconspicuous species of *Leptocyon* throughout the Oligocene and early Miocene. By the middle to late Oligocene (around 30 to 28 Ma), the combined diversity of canids had reached its peak of about 25 species in western North America (figure 6.6). Such a high taxonomic diversity within a single family of carnivorans on a continent was unprecedented before this time and has not been seen since, and it demonstrates the

impact of early canids in the carnivoran communities of North America. This peak diversity was also the first by a carnivoran family in a time otherwise dominated by archaic predators, such as hyaenodonts, in most parts of the world. It signaled that the time for carnivorans had arrived. Indeed, hyaenodonts and other archaic predators had begun to decline, and their roles were eventually taken over by members of the Carnivora.

The Mid-Miocene Climatic Optimum and the Second Diversification of Canids

After the initial shock of plummeting global temperatures in the early Oligocene, temperatures began to rebound modestly, but were largely stable through much of the Oligocene and into the middle Miocene (about 30 to 15 Ma). World climates during this time were still rather mild in modern terms. This stable period, with a gradual warming trend, peaked at the so-called Mid-Miocene Climatic Optimum. Although Antarctic ice first had formed at the beginning of the Oligocene (34 Ma), the ice caps during that epoch were small, located mostly on the high Antarctic plateau, and highly dynamic (fluctuating with time). By the end of the middle Miocene (14 Ma), another drastic drop in temperature occurred, this time with the formation of a permanent Antarctic ice sheet. This fall in temperature precipitated a massive response from plant communities.

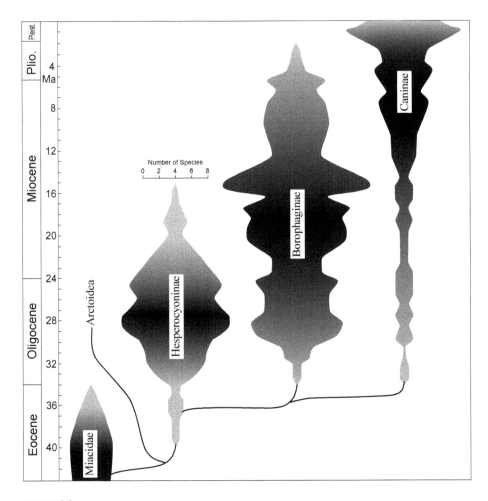

FIGURE 6.6

Diversity of species through time in the three subfamilies of Canidae

The width of each subfamily at a particular time slice in the figure corresponds to the number of species within the subfamily (shown in the scale above the Hesperocyoninae). Although both the Hesperocyoninae and the Borophaginae had a substantial diversity during the Oligocene, members of these subfamilies occupied different niches. The early borophagines were predominantly hypocarnivores, which complemented the hesperocyonines' hypercarnivory. By the time the Hesperocyoninae underwent its drastic decline in the early Miocene, the borophagines began to occupy the hypercarnivorous niches left open by the disappearing hesperocyonines. A similar complementary relationship also existed later between the borophagines and canines.

Direct records of the plant communities in the Miocene are relatively rare. Existing evidence, however, indicates the presence of a certain amount of open grassland, perhaps alongside forests, in the midlatitudes of North America during the early Miocene. During these dramatic climatic changes, mammalian herbivore communities in the continental interiors underwent a similarly drastic turnover. In part because of evolution and in part, for the first time, because of immigration by Eurasian natives, herbivore diversity steadily increased through the early to middle Miocene until reaching an all-time peak in the Mid-Miocene Climatic Optimum (around 15 Ma). This peak is revealed in the extraordinary fossil record of North American horses, a celebrated example of evolutionary adaptation for open vegetation communities. As many as 16 genera (probably including many more species) of horses were present in the middle Miocene, a peak diversity that has steadily declined ever since (figure 6.7). Although lacking a correspondingly rich fossil record of plants, we can reasonably deduce that a high diversity in mammalian herbivores probably corresponded to a high diversity in the plant community as well.

Perhaps not coincidentally, canids experienced their second spurt of diversification in the middle Miocene, although it involved mostly borophagines. (Hesperocyonines were on their way to becoming extinct, and canines were still relatively inconspicuous.) This peak had a somewhat lower taxonomic diversity (20 species) than that of the late Oligocene (25 species), but canids nevertheless experienced their maximum ecological breadth during this time. Canids had by this time acquired more complete hypercarnivorous or hypocarnivorous morphologies, corresponding to a wide range of diets from pure meat to a mix of foods, including a large component of fruits and vegetable matter. In an indirect way, such a high ecological diversity by a family of carnivorans was also a reflection of variable plant communities and perhaps also of competition (or lack thereof) from other groups of carnivorans in North America.

By the late Miocene (8 to 7 Ma), apparently in response to the presence of the permanent Antarctic ice sheet, open grasslands became dominant in midlatitude western North America. Furthermore, near the end of the Miocene, there was a worldwide turnover of the plant communities in the midlatitudes: from C_3-photosynthesizing plants adapted to a high level of atmospheric CO_2 to C_4-photosynthesizing plants adapted to a reduced level of atmospheric CO_2. C_3 plants have a three-carbon photosynthetic pathway and include trees, shrubs, forbs, and cool-season grasses. In contrast, C_4 plants have a four-carbon photosynthetic pathway and include mostly grasses that are better adapted to hot, high-light, and water-stressed environments. C_4 plants seem to have a competitive advantage over C_3 plants in lower levels of CO_2. The transition from C_3 to C_4 plants in the late Miocene was thus another major indication of further climatic deterioration.

In addition to changes in the photosynthetic pathways of plant communities in a lower level of atmospheric CO_2, numerous mammalian herbivores had independently evolved high-crowned cheek teeth by the late Miocene. The height of the crown in herbivores' teeth is defined by the total height of the enamel, including parts that are buried deep in the jaws but that will eventually emerge from the jaws when the crown surface is ground down by use. The high-crowned (also known as

FIGURE 6.7

Comparison of size among mammals from the Miocene (Barstovian) of North America

From left to right: the borophagine dog *Aelurodon ferox*, the small ruminant *Blastomeyx gemmifer*, the hipparionine horse *Neohipparion coloradense*, the ruminant *Cranioceras skinneri*, and the small hipparionine *Pseudhipparion retrusum*, from around 12 Ma. Reconstructed shoulder height of *Neohipparion*: 110 cm.

hypsodont) cheek tooth is defined by a crown height greater than its occlusal length, and it is an ecologically important indicator of diets because it can sustain much more wear associated with the eating of fibrous and abrasive vegetation such as dry grasses. Grasses generally grow in open and seasonally dry environments. Mammals with low-crowned teeth generally eat tender tree leaves and low shrubs and are commonly referred to as *browsers*. By contrast, mammals with high-crowned teeth can handle tougher and drier grasses that are closer to the ground and thus contain more grit that helps wear the teeth faster. These grass eaters are referred to as *grazers*. Late Miocene grazers faced the prospect of eating low-quality vegetation (dry grasses) in greater quantity to compensate for its low nutrition. The high content of grit (wind-blown sands attached to grasses) and the presence of phytoliths (microscopic particles of silica contained in leaf epidermis as a self-defense mechanism) in

many grasses caused additional tooth wear. The combination of these factors was a powerful inducement in the evolution of hypsodont teeth.

As the landscapes opened up, it became increasingly critical for the grazing herbivores to be more cursorial (figure 6.8). A cursorial grazer can better escape from predators and has the added advantage of covering larger grazing range in order to avail itself of optimal pastures. Again, the classic story of the evolution of horses serves as a perfect example. One of the most common ways to develop the ability to run faster and walk for longer distances in a more open landscape is for the limbs to be elongated and lightened, particularly the more distal limb segments (those toward the hands and feet rather than those in the upper arms and thighs). This process is perfectly illustrated in horses by the elongation of their finger and toe bones and the loss of their lateral digits until they eventually stood on a single finger/toe in an ultimate unguligrade posture, as seen in modern horses (Miocene horses did not reach this stage and commonly had three toes).

Against the background of rapid evolutionary change among mammalian herbivores during the late Miocene, canids had to adjust to keep up with their prey. Although carnivorans probably did not care much what their prey ate, they did have

FIGURE 6.8

A scene in Florida during the late Miocene (Clarendonian)

From left to right: the canid *Epicyon haydeni*, the rhinocerotid *Teleoceras fossiger*, the barbourofelid *Barbourofelis loveorum*, the three-toed horse *Neohipparion leptode*, the camelid *Aepycamelus major*, and the protoceratid *Synthetoceras tricornatus*, from around 9 Ma.

to adapt to the increasingly cursorial nature of the herbivore communities. The result is the ever-present saga of the coevolution of predator and prey. It is particularly pronounced among large, pursuit predators such as *Aelurodon*, *Epicyon*, and related forms of borophagines (figure 6.9). The late Miocene canid evolution also marked a decline in diversity and ecological breadth in the borophagines. Competition from new immigrants—such as felids, false saber-toothed cats, large mustelids, and giant ursids—intensified, leading to adaptations for bone cracking among borophagines such as *Epicyon* and early *Borophagus*. Bone cracking as a scavenging strategy was also facilitated by a more open environment, which permitted greater visibility of carcasses and thus more efficient use of leftover skeletons.

The expansion of open grassland in North America in the late Miocene seems also to have set the scene for an early diversification and dispersal of the canines. For much of the time since the origin of the subfamily Caninae at the beginning of the Oligocene (about 33 Ma), the subfamily line of descent was sustained mainly by a few species of *Leptocyon* that were no match for the contemporaneous and more formidable hesperocyonines and borophagines. One feature of *Leptocyon* that distinguished it from the canids of the other two subfamilies was a slender and elongated limb, which, with the loss of a functional thumb and big toe, became a critical advantage when the landscape opened up. By the late Miocene, the early precursors of modern foxes (tribe Vulpini) and an ancestral genus of modern canines (tribe Canini), *Eucyon*, had emerged. In addition, other canids were no longer

FIGURE 6.9

Comparison of size among mammals from the late Miocene (Clarendonian) of North America

From left to right: the borophagine dog *Epicyon haydeni*, the camelid *Aepycamelus*, the ruminant *Synthetoceras tricornatus*, and the hipparionine *Neohipparion*, from around 9 Ma. As shown by this comparison, *Epicyon* was large enough as an individual to hunt some of the medium-size ungulates, although prey the size of the *Neohipparion* and larger would have required group action. Total height to the head of *Aepycamelus*: 3 m.

competing for the foxlike niches for the first time in their long but inconspicuous early history—borophagines had become ever larger and more powerful hypercarnivores and no longer vied for a more generalist predator's way of life.

The time for the subfamily Caninae had finally arrived. One of the most important events in the subfamily's history is its breaking out of North America, the continent that had been the cradle of the family Canidae's evolution for more than three-quarters of its existence and from which the first two of the three subfamilies (Hesperocyoninae and Borophaginae) could not escape (except for a single incidence in the middle Miocene [chapter 3]). A species of *Eucyon* first made it to Eurasia in the late Miocene, setting the scene for the canines' ultimate triumph in the world (chapter 7).

The Pliocene Upheaval and the Great Canid Expansion to the World

Through much of the Pliocene (5 to 1.8 Ma), global climates steadily deteriorated with a precipitous drop in temperatures. By the late Pliocene (about 3 Ma), an arctic ice sheet had begun to develop. This time, ice growth once again coincided with another important tectonic event, the formation of the Isthmus of Panama, which blocked direct ocean circulation between the Atlantic and the Pacific, although global factors probably contributed to the arctic ice formation as well.

Continuing the trend that had started in the late Miocene (6 Ma), the expansion of C4 grasslands was the main theme for plant communities, particularly in the midlatitudes of the world, in an increasingly open landscape caused by a drier, colder, and more seasonal climate during the Pliocene. Mammalian herbivores adapted by acquiring increasingly high-crowned cheek teeth to tackle the tougher and more seasonal grasses. Their limbs also became more cursorial, continuing the earlier trend of lengthening in the distal segments of limbs and reduction in the number of toes. By the Pliocene, the *Dinohippus–Equus* lineage had emerged, and it achieved the ultimate reduction in the number of toes—it retained only the single middle digit, with the second and fourth digits on either side reduced to no more than small remnants of their former sizes. The single-digit horses outcompeted their three-toed horse ancestors and within a short period became the dominant horses in North America. By the late Pliocene to the early Pleistocene, *Equus* immigrated to Eurasia and South America, along with the camels, which, like canids, had been confined to North America during much of their existence.

Borophagines were down to one or two species of *Borophagus* during the Pliocene. *Borophagus* was a highly specialized bone-cracking dog capable of consuming bones to make efficient use of carcasses. The more open landscape probably also aided in better visibility of carcasses left over by other carnivores, but the increasingly cursorial herbivores proved more and more difficult to catch for *Borophagus*, whose skeleton was not adapted for great speed. Low diversity, high specialization (such as bone cracking), and giant size are often signs of terminal lineages in carnivorans. Indeed, the great borophagine subfamily became extinct by the latest Pliocene (2 Ma), when its last species, *B. diversidens*, disappeared from the fossil record.

The subfamily Caninae, however, thrived among the newly emerging, highly mobile herbivore communities. As noted previously, the canines started with longer and more slender limbs, characters that gave them a significant advantage when dealing with increasingly swift prey.

In a wider perspective, the Pliocene was a critical period of expansion for canids. The combination of the most mobile canids ever evolved (in the form of early vulpines and canines), easy access between continents (between North and South America, between North America and Eurasia, and later between Eurasia and Africa via western Europe), and more availability of open grassland created optimum conditions for the greatest expansion of canid distribution ever. During the early Pliocene, canids finally became established in the Old World. Early records of canids in Asia, Europe, and Africa show that they lived in these areas almost simulta-neously. Their first appearances on these continents differs by at most 1 to 2 million years. In Asia, canids arrived in the Yushe basin in the early Pliocene (around 5 Ma). In Europe, the record may be slightly older, as suggested by the presence of *"Canis" cipio* in the latest Miocene (7 Ma) of Spain (Crusafont-Pairó 1950). In Africa, the first appearance of canids is recorded by the fossil of a small fox (*Vulpes riffautae*) from the Djurab Desert in northwestern Chad (7 Ma) (Bonis et al. 2007). As canids arrived on these continents, they quickly diversified in their new homes. This is the third and last peak of canid diversification, which has continued through the Pleis-tocene until today (chapter 7).

The great canid expansion also brought into contact Old World hyaenids and New World canids, two families that are probably the most comparable ecologically (chapter 5). However, by the Pliocene, the competitive landscape had changed sig-nificantly, such that members of the two families were not direct competitors. The newly arrived foxes and jackal-like canids were much smaller than most hyaenids, which by now were large, bone-cracking hypercarnivores. North American canids, those that remained in the New World, also had a limited chance to encounter a highly derived hyaenid, *Chasmaporthetes*, which was the only hyaenid to make it to the New World (figure 6.10). *Chasmaporthetes* may have competed with the last species of borophagines, *Borophagus diversidens*; whereas the former was better at running with a more cursorial limb, the latter was better at cracking bones. If judged by the fossil record, *Borophagus* seems to have outnumbered *Chasmaporthetes* and thus to have made a greater contribution to the carnivoran community.

FIGURE 6.10

Chasmaporthetes

The Pliocene hyaenid *Chasmaporthetes*. Reconstructed shoulder height: 80 cm.

By about 3 Ma in the Pliocene, the formation of the Isthmus of Panama connected the North and South American continents, which had been isolated from each other for millions of years, since the end of the Cretaceous (65 Ma). This connection resulted in the Great American Biotic Interchange, in which numerous land mammals on either continent were able to cross the land bridge and became part of the fauna on the adjacent continent. Carnivorans that immigrated to South America generally outcompeted native predators (such as borhyaenid marsupials) and quickly established themselves as the dominant components of the predatory communities.

Canids were certainly players in this success story. Starting with only a few lineages of canines in the Pliocene of Central America and southern North American, they experienced an explosive radiation once they arrived in South America. Today, the South American canids are the most diverse group of canids on any continent. The 11 species of South American canids constitute almost one-third of the entire canid diversity and are the largest group of predators in South America among various families of living Carnivora (see appendix).

The Pleistocene Ice Age and the Establishment of Modern Canids

The last epoch in the Cenozoic is the Pleistocene (1.8 to 0.01 Ma). Often called the Ice Age, this epoch is marked by extensive continental ice sheets that repeatedly advanced toward the midlatitudes. Temperatures dropped to the lowest levels ever in the Cenozoic, and during the peak glaciations continental ice sheets up to 3,000 m thick (nearly 2 miles) advanced as far south as Nebraska, Illinois, and Kansas in North America, as well as over Scandinavia and much of northern Europe, occupying about one-third of the globe. When climates shifted from warm and humid to cold and dry, and vice versa (during the interglacial periods), the plant communities underwent drastic, cyclical changes, often within short periods of time. During the glacial maxima, distributions of animal and plant communities suffered from contractions toward the equator; during the interglacial periods, they reoccupied formerly lost ground toward the higher latitudes.

To cope with the extreme coldness during glacial maxima, many large mammalian species, particularly herbivores, attained giant sizes in accordance with the general biological rule that animals in colder climates tend to increase their body sizes to help conserve body heat and to store larger quantities of fat to cope with winter weather. As a result, Pleistocene megafauna emerged in the northern continents (North America and Eurasia). Such giants as woolly mammoths, buffalo, giant deer, and woolly rhinos roamed in Eurasia; mammoths, mastodonts, giant ground sloths, large saber-toothed cats, and dire wolves reigned supreme in North America. Most members of these megafauna, especially those from North America, became extinct at the end of the Pleistocene. Interestingly, the gray wolf (*Canis lupus*) was one of the few exceptions and is still one of the most successful large canids in the world. If one counts the domestic dog as a highly specialized adaptation for cohabiting with humans, as some people have claimed, then *Canis* has achieved its ultimate success in occupying nearly every corner of the world.

Canids had lived in the harshest climates since the beginning of the Cenozoic, and Pleistocene canids were probably little affected by climatic conditions, judging from the existence of modern arctic wolves and foxes, which, along with polar bears, are the hardiest carnivorans in the Arctic. In fact, much of the gray wolf evolution apparently played out in high-latitude or even circum-Arctic regions. Given the worldwide distribution of canids during the Pleistocene and the diverse environments in which they lived, it is difficult to detect general patterns of canid evolution associated with changes in environment or in herbivore communities, as happened more obviously in the late Miocene. In general, large canines, such as the genus *Canis* and related genera such as *Cuon* and *Lycaon*, tended to become larger and more hypercarnivorous, a trend possibly accelerated by periodic glaciations, but perhaps also simply by their own tendency to become larger as their body size passed a certain threshold, as also happened to the earlier hesperocyonines and borophagines, which did not experience glaciations (chapter 5).

In Europe, the beginning of the Pleistocene was marked by the so-called Wolf Event, the appearance of wolflike *Canis* near the Pliocene–Pleistocene boundary (1.8 Ma). Since then, *Canis* has had a continuous presence in Eurasia, along with various species of foxes and raccoon dogs. Early humans, both *Homo erectus* and *H. sapiens*, must have had close encounters with canids because the hunter-gatherer lifestyle is broadly similar to the canid lifestyle. Ice Age humans thus may have competed with some larger species of canids. By the latest Pleistocene, this close encounter resulted in the domestication of the first canid in the Middle East or Europe or possibly China (chapter 8).

In North and South America, canids played an important role as top predators in the megafauna. The best-known example is the dire wolf (*Canis dirus*), which by this time had reached 68 kg in body weight, not exceeded by any other canine before or since (the only canids larger than the dire wolves were advanced species of *Aelurodon* and *Epicyon*). Fossil records of the dire wolf are widespread in much of the United States and Mexico south of the last glacial ice cap. Dire wolves also ranged into Andean South America. In the southern California locality of Rancho la Brea, they, along with the saber-toothed cat (*Smilodon*), occurred in high numbers relative to other carnivorans. Blaire Van Valkenburgh and Fritz Hertel's (1993) study of canine tooth breakage on the La Brea dire wolves suggests that by the late Pleistocene, dire wolves used carcasses more systematically than is seen in modern wolves,

FIGURE 6.11

Cynotherium sardous

Life reconstruction of *Cynotherium sardous*, an extinct canid from the late Pleistocene (around 15,000 years ago) of Sardinia. Shoulder height: 44 cm.

and they were probably in intense competition with other large predators, such as saber-toothed cats and lions, for limited resources. Along with the megafauna, dire wolves became extinct by the end of the Pleistocene.

The frequent advances and retreats of continental ice sheets caused correspondingly drastic rises and falls of sea levels. Massive amounts of seawater were stored in the ice sheets, and sea levels dropped by as much as 120 m during the glacial maximum. A direct consequence of this lowering of sea levels was the emergence of connections between the mainland and some of the surrounding islands, and then the reisolation of the islands when sea surfaces returned to higher levels. Such a repeated connection and disconnection between mainland and islands created opportunities for mainland animals to invade islands, but then to be subsequently cut off from their mainland relatives. Isolation on an island is often a direct impetus for speciation, commonly through adaptations to island conditions and a lack of genetic exchange with mainland populations. Two such examples of canid speciation are the extant island gray fox (*Urocyon littoralis*) from the Channel Islands, off the shore of southern California, and the extinct Pleistocene *Cynotherium sardous* from the island of Sardinia, off the west coast of Italy (figure 6.11). The former is a sister species of the mainland gray fox (*Urocyon cinereoargenteus*), and the latter may be derived from a hypercarnivorous, wolflike *Xenocyon* (Lyras et al. 2006). Finally, the Falkland Island fox (*Dusicyon australis*), which was hunted to extinction in the nineteenth century, may also be a product of speciation as a result of connection with the South American mainland and then isolation afterward, although the great distance between the Falkland Islands and Argentina makes such a connection more difficult to imagine, and some measure of overwater rafting may have been involved for the fox to arrive on the islands.

Finally, the modern Holocene epoch (the previous 10,000 years) is marked by a rebound of temperatures and retreat of polar ice caps, creating the environmental conditions that we live in today. Pleistocene canids on the various continents, particularly those in the late Pleistocene, are increasingly similar to their living counterparts. Probably all extant canine species evolved during the late Pleistocene, as is the case for most, if not all, extant carnivorans.

In figure 6.6, we chart the changes in diversity through time to compare the histories of the three canid subfamilies. The figure reveals a relay race among these groups since their origin in the early Cenozoic (around 40 Ma) of North America. Each subfamily shows an increase in diversity following the extinction of the preceding lineage. In the Caninae, the increase did not happen until the end of the Miocene, with maximum diversity not established until the Pliocene. Interestingly, the first departure from the generalized mesocarnivorous adaptation was to hypocarnivory, a broadening to a mixed diet. This adaptation is still represented by the gray fox (*Urocyon*) of the New World (whose fossil sister lineage was *Metalopex*), the raccoon dog (*Nyctereutes*) of eastern Asia, and the crab-eating zorro (*Cerdocyon*) of South America, which is related to Pliocene species in North America. The Borophaginae had a similar adaptive history while it was represented by small species. The relationship between the established group and the new group is suggestive of competition until extinction of the older group reduces competitors, allowing diversity to increase markedly in the younger group. Figure 6.6 suggests that the strong differentiation in the Caninae over the past 4 million years was promoted by the loss of the Borophaginae. A similar relationship is indicated by the history of the Hesperocyoninae and its effect on the Borophaginae.

Another feature of this successive aspect of canid evolution is that the adaptation of hypercarnivory is restricted mostly to lineages in the later part of each subfamily's history, and the adaptation seems to be related mostly to the increasing size of both predators and prey. Of course, this generalization applies most clearly to the Borophaginae and the Caninae. The Hesperocyoninae was primitively hypercarnivorous and developed the molariform lower carnassial only in its latest genus, *Osbornodon*, which perhaps increased its longevity as a group until comparable large borophagines (for example, *Tomarctus*) came into existence in the middle of the Miocene (16 Ma).

7 GOING PLACES
BRAVING NEW WORLDS

AS PREDATORS WELL SUITED FOR TRAVELING long distances, modern canids are the only family of Carnivora to have a truly worldwide distribution (except Antarctica). This wide dispersal has in no small part been due to their ability to expand their home ranges and to achieve long-distance dispersal across continents and habitats. Canid zoogeography provides insights into the intricate relationships among canid species around the world. The immigrations of major lineages can be traced across different continents during their geologic history. Studies of ancestral–descendant (phylogenetic) relationships in the fossil record and the history of continental reconfigurations (plate tectonics) allow us to draw some conclusions about the timing, direction, and identity of various dispersals during the canids' history.

Early Endemism

More than two-thirds of the history of the canids was played out in North America, the family's continent of origin. With the exception of a single hesperocyonine lineage, two of the three subfamilies, the hesperocyonines and the borophagines, never left their home continent. Almost from the very beginning of their history, canids seem to have had ample opportunities to cross Beringia, a periodic land bridge between eastern Siberia and Alaska (the present-day Bering Strait), to expand to the Old World. Since the late Eocene (40 Ma), numerous carnivorans did just that, including members of the Mustelidae, the Procyonidae, the Ursidae, the Felidae, and the Nimravidae (an archaic group of saber-toothed "cats"), many of which crossed more than once and in both directions.

Despite these intercontinental immigrations, mammalian communities maintained a large measure of identity on their home continents. Land bridges, such as Beringia and the Isthmus of Panama, tended to be intermittent and were controlled by global eustatic sea-level changes due to ice sheet formation at the poles or local tectonic reconfigurations of adjacent plates. Land bridges located in high latitudes or near the equator, such as Beringia and the Isthmus of Panama, had a strong filtering effect in that specific environments in and around the land bridges encouraged the passage of some species, but prevented others from going through. This is the reason why fauna in North America always maintained a distinct identity through time despite the occasional connections with the outside world (either Eurasia or South America). Organisms that are confined to a particular region or continent are called *endemics*. Several major groups of North American mammals were either entirely or largely endemic in their continent of origin, including the oreodonts (extinct even-hoofed mammals), the camelids (camels, lamas, and their extinct relatives), and the equids (horses). Among carnivorans, the hesperocyonines and borophagines were the best examples of two groups of endemic predators that played a critical role in the North American predatory landscape for many millions of years during the middle to late Cenozoic (30 to 8 Ma) (figure 7.1).

Such endemism in both herbivores and carnivores during much of the middle Cenozoic was apparently the rule rather than the exception. The pattern of endemic distribution may reflect the effectiveness of isolation mechanisms (such as barriers formed by the seas between North America and Eurasia, and between North America and South America) or the filtering effect of a land connection that acted as an environmental bottleneck to prevent easy crossing.

Another factor that contributes to our assumption of this apparent endemism may be related to our highly biased fossil records. Most of the known fossil records are concentrated in the midlatitudes of the northern continents (Eurasia and North America). Although such an aggregation of finds may be partially attributed to socioeconomic factors (industrialized countries in the midlatitudes of the northern continents tend to explore their fossil records more fully than other countries do), this midlatitude bias may also be due to the actual preservation of fossils.

Terrestrial fossils of Cenozoic mammals tend to be more commonly preserved in fluvial (river), lacustrine (lake), and floodplain deposits. Tropical forest environments do not readily preserve fossils, or if they do, the sediments that trap the fossils are often poorly exposed because of the lush vegetation coverage. In high latitudes, in contrast, the continental ice sheets during the Ice Age (Pleistocene [1.8 to 0.01 Ma]) scraped away many of the relatively soft sediments on top that were deposited in the late Cenozoic. The result is a dearth of knowledge about fossils from both high and low latitudes, leaving us a biased view of the middle (see figure 2.14). Therefore, we know preciously little about the high latitude faunas, which would be the most informative for immigration events across Eurasia and North America.

Nonetheless, paleontologists do occasionally stumble on rare events of animal migrations. Until recently, hesperocyonines were known only in North America,

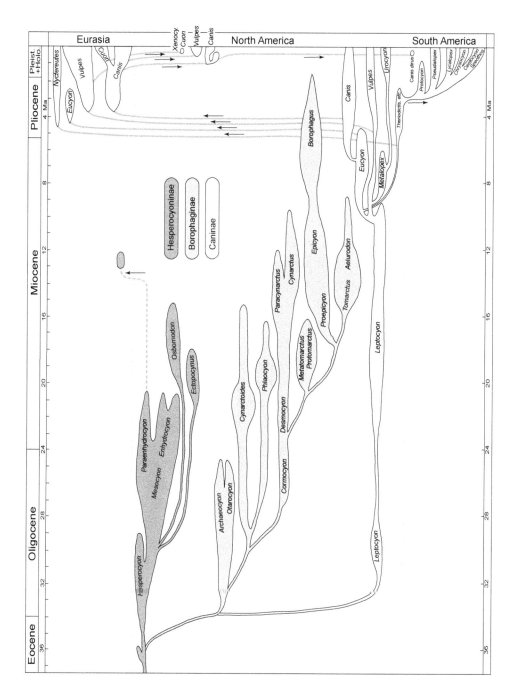

FIGURE 7.1

Phyletic relationships and intercontinental migrations of canid genera

Only selected genera are listed, and the actual diversity in each subfamily (Hesperocyoninae, Borophaginae, and Caninae) is somewhat higher than shown. The width of each lineage in the figure roughly corresponds to the diversity of the lineage through time. For a more realistic sense of diversity through time, see figure 6.6. Arrows indicate directions of migrations.

which was puzzling to paleontologists. Why didn't the hesperocyonines make it to Eurasia? Were they not well adapted to roam large areas, as are their living counterparts, such as the wolves and foxes? Were they confined to the middle latitude of North America and unable to cross the Beringian land bridge in the high arctic region? A tantalizing hint emerged in the summer of 2005 when a team of Chinese paleontologists (led by Xiaoming Wang) unexpectedly found a partial skull of a medium-size hesperocyonine in the middle Miocene beds (Tunggur Formation, around 13 to 12 Ma) of the Inner Mongolia Province in northern China. It is the oldest canid fossil to be discovered anywhere outside North America. Despite this single hesperocyonine in Asia, however, it is fair to say that hesperocyonines did not significantly affect the Old World predator community.

The record of borophagine dispersal is even more dismal. Known fossil records of the borophagines occur as far north as the northern United States and as far south as Honduras and El Salvador. Despite their successes in both diversity and abundance, the borophagines apparently never left North America. The rise of large, hypercarnivorous borophagines apparently followed the maximum diversification of horses in the middle to late Miocene (18 to 10 Ma). It is thus reasonable to assume that horses probably constituted a major source of prey for the borophagines. Why, then, didn't the borophagines follow the *Hipparion* (three-toed) horses to the Old World? The dispersal of *Hipparion* to Eurasia in the late Miocene (about 12 to 11 Ma) represents a major event in the history of mammalian zoogeography. Upon their arrival in Eurasia, the three-toed horses quickly diversified and expanded to the entire Old World to become a key faunal component wherever they went. They became ubiquitous in the late Cenozoic (11 to 5 Ma), and their impact on the ungulate communities must have been high.

Although for want of fossil evidence we are unable to fathom the cause of the borophagines' failure to disperse, we can take a look at the competitive landscape as a way to explore possible scenarios. Canids and hyaenids independently evolved many similar features. So convergent were the predatory behaviors of members of these families that competitive exclusion may have played a role in explaining why canids and hyaenids were confined to their continents of origin during much of their histories. Throughout these histories, there were numerous opportunities for canids and hyaenids to disperse to continents other than their own—many other carnivorans, including felids, made the trip several times—and canids and hyaenids, with their highly developed cursorial adaptations that enabled them to travel long distances with ease, were ideally suited to cross the continents. Yet neither family made a significant dent in the other's "home turf." A single hyaenid, *Chasmaporthetes*, arrived in North America in the Pliocene (4 Ma), but was never successful enough to be a significant component of the New World's carnivoran fauna (see figure 6.10). *Chasmaporthetes* was probably in direct competition with *Borophagus*, which was well into the last leg of its journey toward extinction. It is conceivable that some borophagines may have made it to the Old World, but met strong competition from hyaenids and were quickly eliminated.

An alternative scenario may be that the environments in and around Beringia had a strong filtering effect on which species could cross the land bridge. For example, a heavily forested environment would have favored predatory species that preferred wooded areas. Indeed, many of the medium to large carnivorans that successfully immigrated to North America during the Miocene were felids and ursids, two families that tend to prefer the cover afforded by trees (early hemicyonine ursids that came to North America, however, had long, digitigrade legs that were probably adapted to running in open environments). Recent discoveries of late Miocene to early Pliocene (6 to 4 Ma) red pandas and meline badgers (Old World badgers) in eastern Tennessee, which is part of the Eastern Deciduous Forest of the United States, may even suggest the presence of forest corridors across much of Beringia. If so, Beringia may have served as an environmental bottleneck to limit the dispersal of borophagines, which were adapted to open land. Unfortunately, we have preciously few late Cenozoic fossil records in northern Siberia and Alaska, where past faunal exchange must have occurred, and until such a deficiency can be remedied, history's apparent paradox will remain a mystery.

Dispersal of Caninae

The subfamily Caninae, like its predecessors, remained landlocked in North America for more than two-thirds of its history, during which *Leptocyon* was living in the shadow of both hesperocyonines and borophagines. Canines, however, did eventually achieve a breakthrough by the late Miocene (7 Ma), making appearances in Europe, Africa, and Asia in short succession. One of the Caninae's distinguishing characters is its more advanced stage of locomotive adaptation. For example, compared with the more primitive *Leptocyon*, early *Vulpes* had lengthened legs, a reduced first digit in the metatarsals, and no epicondylar foramen on the distal humerus—characters that are commonly associated with lengthening of the stride and reduction in the weight of the limbs (chapter 4). These cursorial developments are certainly more advanced relative to developments in the hesperocyonines and borophagines. However, it is not clear if such characters alone were enough to permit the canines to break through the geographic barriers or if environmental factors—that is, the opening up of the landscapes (chapter 6)—may have contributed to the eventual dispersal of canines (figure 7.2; see figure 7.1). Whatever the reason that canines were able to immigrate to the Old World, the predatory community there was never the same after they arrived.

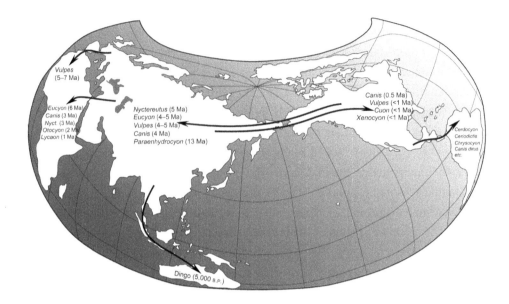

FIGURE 7.2

Intercontinental migrations of canids

As one of the most mobile groups of carnivorans, canids have worldwide distribution as a result of multiple migrations across continents. The lists given here of canids that arrived at each continent are not exhaustive, and the canids listed are a subset of all canids that made it to that particular continent. Numbers following each taxon approximate the time of arrival millions of years ago (Ma).

EUCYON AND OTHER EARLY CANINES IN THE OLD WORLD

The first canine to arrive in the Old World was a coyote-size form named *"Canis" cipio* (Crusafont-Pairó 1950). A maxillary fragment with P3 to M2 and an isolated lower carnassial (m1) from the Teruel basin of central Spain in deposits of Turolian age (late Miocene [about 8 to 7 Ma]) offer all that we know about this species. The relatively primitive morphology of the teeth suggests that this species was either *Eucyon* or a very primitive *Canis*. However, lacking more complete materials, we cannot determine its precise phylogenetic position.

Lorenzo Rook (1992) reported a new species of *Eucyon*, *E. monticinensis*, from the Monticino gypsum quarry near the village of Brisighella in Italy and from Venta del Moro in Spain. These *Eucyon* materials are also late Miocene in age (about 7 to 5 Ma), but probably slightly younger than *"Canis" cipio*.

In Africa, the relatively poor canine fossil records have traditionally indicated a rather late arrival of canines on the continent. However, recent findings (often associated with the explosive new knowledge of hominid-bearing localities) have pushed the canine records to ever earlier ages, comparable to the European records.

Jorge Morales, Martin Pickford, and Dolores Soria (2005) described a new species of *Eucyon, E. intrepidus*, from the Lukeino Formation (6.1 to 5.7 Ma) in the foothills of the Tugen Hills in the Baringo District, along the Great Rift Valley in western Kenya. This species is morphologically similar to early *Eucyon* species (but smaller in size) from the late Miocene of Europe and early Pliocene of Asia (China).

In another recent finding associated with the earliest hominid remains (*Sahelanthropus tchadensis*), Louis de Bonis and his co-workers (2007) report a new fox, *Vulpes riffautae*, from the Djurab Desert, northwest of N'Djamena in the West African country of Chad. This new fox is associated with fossils from the late Miocene (around 7 Ma). If this age estimate is correct, then *Vulpes riffautae* would be the earliest vulpine record in Africa, possibly even the earliest in the entire Old World. The Chad fox was very small, barely larger than a modern fennec (*Vulpes zerda*), which is the smallest living fox in the world.

Based on these materials, it appears that canines of the *Eucyon–Canis* stage of development and a small fox were already present in the late Miocene of Europe and Africa. Fossil records of canines in Asia, however, are slightly younger, from the early Pliocene and later (less than 5 Ma), even though canines must have passed through the vast stretch of Asia to arrive in Europe and Africa. Such a lag in the arrival of canines in Asia may simply be attributed to relatively poor documentation of the Asian fossil record. However, an alternative scenario may be that the first canines that crossed over Beringia stayed in higher latitudes, bypassing the midlatitude regions of Asia, where our records are currently based, and going straight to Europe.

Whatever the scenario, canines are firmly recorded in the early Pliocene deposits of northern China, particularly in the Yushe basin in central Shanxi Province. In the Yushe basin, *Nyctereutes* (*N. tingi*) appeared shortly after 5 Ma, followed by *Eucyon* (*E. zhoui* and *E. davisi*) slightly later. In fact, the presence of *Eucyon* in the Old World was first documented by Richard Tedford and Qiu Zhanxiang (1996) from the Yushe records. Since then, increasing evidence suggests that this transitional genus had a modest success outside its continent of origin, North America. Several of its species are currently recognized in the fossil records of Asia, Europe, and Africa, and these species are in general considered the earliest canines in the Old World.

RACCOON DOG

The raccoon dog (*Nyctereutes*) was an important early immigrant to the Old World. In the Yushe basin, the primitive species *N. tingi* entered the record in the latest Miocene to the middle Pliocene (5.5 to 3 Ma). At about the same time, the European species *N. donnezani* lived in the area of present-day Perpignan, France. These ancestral raccoon dogs were coyote-size animals, much larger than their modern

FIGURE 7.3

Nyctereutes terblanchei

Reconstructed skull and head of *Nyctereutes terblanchei*. The mandible reconstruction is based on a fossil from Kromdraai-A (1.6 Ma); the skull, after other species of the genus. Although the modern raccoon dog has an original Asian distribution with artificial introductions into Europe, the distribution of *Nyctereutes* in the Pliocene and Pleistocene (5 to 1 Ma) reached as far as South Africa. Mandible length: 14 cm.

counterpart. More advanced species, *N. sinensis* in China and *N. megamastoides* in Europe, soon appeared in Eurasia, and they eventually led to the living species *N. procyonoides* in East Asia. By the early Pliocene (3.8 to 3.5 Ma), *Nyctereutes* had reached Africa, but the African raccoon dog did not survive beyond the Pleistocene (1 Ma) (figure 7.3). The modern raccoon dog is a descendant of the *N. sinensis–megamastoides* lineage in Eurasia, surviving because it became smaller in size and more hypocarnivorous in dental adaptations. The modern East Asian raccoon dog was introduced to northern Europe in the twentieth century.

Although the Old World history of *Nyctereutes* can be traced by a useful fossil record, it is still far from clear what its original stock in North America was. *Nyctereutes* seems to be closely related morphologically to the South American crab-eating fox (*Cerdocyon*) in that both share a somewhat hypocarnivorous dentition and an enlarged angular process (a small bony protrusion on the lower jaw for insertion of the pterygoid muscles), among other characters. According to this scenario, *Nyctereutes* probably arose from a common ancestor of a *Cerdocyon–Nyctereutes* lineage in North America. Records of *Cerdocyon* from the early Pliocene (5 Ma) of Texas and New Mexico suggest that the Old World *Nyctereutes* could have arisen from a North American stock close to *Cerdocyon*. However, molecular studies by Robert K. Wayne and his colleagues (1997) place the raccoon dog at the base of the living canines. This hypothesis suggests that the raccoon dog is either more basal to the vulpines or within the fox lineage (see also Lindblad-Toh et al. 2005).

BAT-EARED FOX

The living African bat-eared fox (*Otocyon megalotis*) is a peculiar canine distinguished by an extra molar at the back of the lower tooth row due to its insectivorous diet. Based on our morphological analysis (Tedford, Taylor, and Wang 1995), *Otocyon* seems to be closely related to the North American gray fox (*Urocyon*), which has never left its native continent. The *Otocyon* lineage had arrived in the Old World by the late Pliocene (3 Ma) in the form of a transitional genus, *Prototocyon*, from India, although its record in Africa may be somewhat earlier (around 4 to 3 Ma). The extant *Otocyon* was presumably derived from *Prototocyon* in Africa.

FOXES

The earliest and most primitive foxes (tribe Vulpini) were a small, California species (*Vulpes kernensis*) and a larger, continent-wide species (*V. stenognathus*) from the late Miocene (about 9 to 5 Ma) of North America. Although a small *Vulpes*, *V. riffautae*, had made it to West Africa by as early as 7 Ma, Eurasian foxes appeared somewhat later. *Vulpes beihaiensis* from the early Pliocene (4 Ma) of China and *V. galaticus* from the early Pliocene of Çalta in Turkey were among the earliest foxes in Asia and Europe. A number of *Vulpes* species appeared in the Plio-Pleistocene of Eurasia and Africa. Up to 12 species of *Vulpes* (including the arctic fox, which is sometimes included in a distinct genus of its own, *Alopex*) have survived to the present time throughout Africa, Eurasia, and North America, making it the most diverse living canine genus.

Vulpes is one of a few canines that returned to North America, where it had originated. After the two ancestral species of North American *Vulpes* in the late Miocene
(6 Ma), *V. kernensis* and *V. stenognathus*, foxes became rather sparse during the
Pliocene (5 to 1.8 Ma), as indicated by the fossil record. The genus did not reappear
in North America until the middle Pleistocene (1 Ma). By then, the main theater
of *Vulpes* diversification had shifted to Eurasia. By the latest Pleistocene, the red
fox (*V. vulpes*) and arctic fox (*V. lagopus*) had expanded to North America. The
swift fox (*V. velox*) and kit fox (*V. macrotis*), however, may have been native North
American species.

CANIS GROUP

The first Old World immigrants—*Vulpes*, *Eucyon*, and *Nyctereutes*—were not conspicuous predators of their time, despite their presence in the late Miocene and early
Pliocene. These small to medium-size canines became important components in the
carnivoran community, but were far from achieving a top-predator status. The arrival of *Canis*, however, finally ushered in the era of dominant canines in the northern continents.

During the middle Pliocene (4 to 3 Ma), a wolf-size *Canis*, *C. chihliensis*, first appeared in northern China. Shortly later, Eurasia became a vast playground for *Canis*
evolution, setting off a wave of diversification that established the family's ultimate
success. *Canis* quickly spread to Europe through the species *C. arnensis*, *C. etruscus*
(figure 7.4), and *C. falconeri*, which either retained primitive characters of the genus
or acquired hypercarnivorous characters in the direction of the hunting dogs (*Cuon*
and *Lycaon*). This sudden expansion of *Canis* at the beginning of the Pleistocene
(about 1.8 Ma) is commonly known as the Wolf Event and is associated with the origin of the mammoth steppe biome following intense continental glaciations.

The gray wolf (*Canis lupus*) appeared in Europe toward the end of the middle
Pleistocene (0.8 Ma), but not in midlatitude North America until the latest Pleistocene (about 0.1 Ma). An older record of the gray wolf can be found in the early to
middle Pleistocene Olyor Fauna of Siberia and its equivalent in Alaska (the arctic
Beringia). Wolves thus originated in Beringia, perhaps in coevolution with the large
ungulate assemblage of the arctic biome. They seem to have invaded the midlatitudes of North America, as did many of the large Beringian ungulates, only during
the last glacial cycle, becoming characteristic members of the latest Pleistocene and
living fauna.

That the modern gray wolf has achieved the ultimate success of being the most
widely distributed species of large carnivorans is probably due to its propensity for
dispersal. One of the strategies that gray wolves use to obtain a breeding position

FIGURE 7.4

Comparison of *Canis etruscus* and *Pachycrocuta brevirrostris*

Life reconstruction of the European dog *Canis etruscus*, from the early Pleistocene (1 Ma), shown to the same scale with the giant hyena *Pachycrocuta brevirrostris*. Large hyaenids and canids sought similar resources (carcasses and live ungulate prey), and they would have competed in the European ecosystems of around 1 Ma.

is *directional dispersal*. Young adult wolves of both sexes, in order to escape their subdominant position and to establish their own breeding pack, move in a single direction for more than 800 km. Such a remarkable ability to disperse for a long distance in a single generation is perhaps the most important reason behind their wide distribution. The modern gray wolf and red fox have the widest distributions of any mammalian species, occupying the entire northern continents of Europe, Asia, and North America (Holarctic), as well as North Africa in the case of the red fox.

The dire wolf (*Canis dirus*), made famous by the stunning number of individuals recovered from the Rancho la Brea tar pits in Los Angeles, is another lineage that can be traced to Eurasia. This wolf of the middle to late Pleistocene (1 to 0.01 Ma) is the largest species ever to have evolved in the subfamily Caninae. Our phylogenetic analysis suggests that *C. dirus* arose from the large wolf *C. armbrusteri*, which suddenly appeared in the early Pleistocene (1.5 Ma) of North America. These species share a suite of hypercarnivorous characters, indicating that the dire wolf arose in North America in the middle Pleistocene (1 Ma). Before its extinction in North America, the dire wolf managed to expand to the north and west coasts of South America and establish itself as the most formidable predator in the New World. It had become extinct by the latest Pleistocene (10,000 years ago) as part of the megafauna extinction.

In a comparable evolutionary burst in the middle Pleistocene of Eurasia, the *Canis falconeri* group gave rise to the widespread hypercarnivore *Xenocyon*, which in turn gave rise to the dhole (*Cuon*) and the African hunting dog (*Lycaon*). *Cuon* appears in the fossil record in the middle Pleistocene deposits of Southeast Asia (*C. javanicus fossilis*). An isolated record of *Lycaon* is found in the late Pleistocene Hayonim Cave of Israel. The *Lycaon* record in Africa is poorly known, but this genus possibly occurred as early as the middle Pleistocene, as indicated in the fossil records at the Elandsfontein site of South Africa (Ewer and Singer 1956). During the late Pleistocene, *Xenocyon* and *Cuon* also wandered into North America, briefly enriching the fauna of the New World.

CERDOCYONINES

Our morphological analysis indicates that the South American canines (subtribe Cerdocyonina) belong to a natural lineage whose common ancestor was a sister group of the *Eucyon–Canis–Lycaon* clade (Tedford, Taylor, and Wang 1995) (see figure 7.1). Such a relationship is also increasingly borne out by DNA sequence analysis (Lindblad-Toh et al. 2005). Long before the establishment of the Isthmus of Panama in the middle Pliocene (about 3 Ma), however, certain lineages with cerdocyonine affinities had already shown up in North America during the late Miocene to early Pliocene. They are represented by records of *Cerdocyon* (*C.* sp. A) from the late Miocene (6 to 5 Ma) and of *Theriodictis* (*T.*? sp. A) and *Chrysocyon* (*C.* sp. A) from the early Pliocene (5 to 4 Ma). Our proposed sister relationship between the South American clade and *Eucyon davisi* also implies an early appearance of the cerdocyonines, possibly a few million years earlier than the current fossil record indicates and presumably in the Central America region where the fossil record is very poor.

This early diversification of the cerdocyonines in North America prior to the opening of the Isthmus of Panama implies that the migration to South America must have been undertaken by more than one lineage. Representatives from at least these three genera—*Cerdocyon*, *Theriodictis*, and *Chrysocyon*—must have independently crossed the Isthmus of Panama, and others probably did as well. Thus the large diversity of South American canines was probably built on top of a group that was modestly diverse while still in North America.

Once in South America, the cerdocyonines suddenly faced a fauna made up of native marsupials and notoungulates (large herbivores restricted to South America). The cerdocyonines underwent an explosive radiation to become the most diverse carnivoran group in South America, presumably at the expense of the native borhyaenids (chapter 2). In addition to forms that constitute the modern cerdocyonines—such as the short-eared dog (*Atelocynus*), the bush dog (*Speothos*), and the various zorros (*Pseudalopex* and *Lycalopex*)—the hypercarnivorous genus *Protocyon*, which is extinct, was also part of the native radiation. Beside the cerdocyonines, species of

Canis were able to take advantage of the land bridge. The oldest canine species, *C. gezi*, is known only from deposits of Ensenadan age (early to middle Pleistocene [1 to 0.5 Ma]) in Argentina. *Canis gezi* and *C. nehringi* found in Lujanian (late Pleistocene [0.2 Ma]) sediments in Argentina possibly are closely enough related to the dire wolf (*C. dirus*) to suggest an earlier Pleistocene arrival and the modest differentiation of a dire wolf lineage in South America. Together, *Cerdocyon* and *Canis* established themselves as a dominant group of predators and permanently altered the carnivore communities in South America.

8 DOMESTIC DOGS

THE STATUS OF DOMESTIC DOGS as "man's best friend" indicates that they evoke more emotional responses than do other animals. Indeed, people often lose their objectivity over questions about dogs. Terms such as *beauty*, *intelligence*, and *loyalty* are frequently ascribed to dogs as though they are qualities inherent to dogs. To understand the relationship between humans and dogs requires, however, that we know more about the natural history of domestic dogs as they relate to their wild ancestors and about the history and process of their domestication.

Why Is Domestication Important?

Although most people have no difficulty relating to dogs as something very special in their lives, probably few of them have contemplated the evolutionary significance of domestic dogs in the history of human civilizations. First of all, dogs were the first domestic animal, a fact that has had profound consequences in human history. Domestic animals are so ubiquitous now that we often take for granted both them and all the conveniences they offer us in our daily lives. As Jared Diamond (1997) has pointed out, however, large domestic animals can be one of the deciding factors in the rise and fall of civilizations. Domestic animals are vital sources of food, labor, transportation, and clothing—all essential items for societies. When civilizations clash, those with the right domestic animals (such as horses in a war) can have an overwhelming advantage that tilts the eventual outcome in their favor.

The very idea of a certain invention is often more important than the way that invention is achieved. Once an idea is born, people find different ways to realize it. In the case of domestication, the very idea that animals can be harnessed to human betterment is much more important than which specific animal to domesticate and in what sequence. Consequently, if dogs are truly the first animal to be domesticated, then they can rightfully claim to have instilled in humans the notion that animals can be raised in human dwellings, a radical idea that has had profound effects on the earth's biota (as in the case of dingoes in Australia), on people's relationships with animals, and on the way people live. Once such a notion became established, domesticating other animals would be a matter of course by choosing the right species and by experimenting with the right techniques for domestication through trial and error.

It is perhaps no coincidence that the first domestication happened to the wolf, a carnivoran with high social intelligence and well adapted to living with fellow beings. Diamond (1997) notes that most large mammals are highly resistant to human domestication. There are only about 14 successful and economically important large domestic mammals, most of them originally from Eurasia and all of them herbivores. None of the 14 large herbivores possesses the kind of social complexity that dogs have and presumably were initially far less domesticable than dogs. Intriguingly, all of the top five domestic mammals in terms of economic importance—sheep, goats, pigs, cows, and horses—were first domesticated in and around the Fertile Crescent in western Asia (a possible exception is the pig, which may have been first domesticated in China). It is perhaps no coincidence that these five domestic mammals are also among the first to show up in archaeological records in western Asia (between 8,000 and 4,000 years ago). One of the earliest records of domestic dogs is from Israel about 12,000 years ago, well before the records of the big five. Could it be that people in the Middle East and western Asia first learned the ideas of domestication through their association with dogs and quickly applied the same idea to other large herbivores? If so, such an idea was nothing short of revolutionary, leading people to work with large, wary herbivores, to harness greater power, to travel longer distances, and to carry more goods than they otherwise would have been able to by their own means.

History of Dog Domestication

In 1977, a father-and-son team, Stanley J. Olsen and John W. Olsen, advanced the thesis that the likely area of origin for domestic dogs was China. They subsequently attempted to tackle systematically the fossil records of domestic dogs at a time when issues about animal domestication were increasingly attracting archaeologists' attention (Olsen 1985). They started by assuming that a small subspecies of wolf would be the most likely candidate, and the modern Chinese and Mongolian subspecies *Canis lupus chanco* seemed to fit the bill. The Olsens then studied the *Canis* materials from the Zhoukoudian sites, about 50 km southwest of the capital city, Beijing, famous for the discovery of Peking Man ("Peking" is the Wade-Giles spelling of

"Beijing"), a middle Pleistocene record of *Homo erectus* (500,000 years ago). The *Canis* in question had been named *Canis lupus variabilis* by Pei Wenzhong in 1934, and Stanley Olsen pointed out that in size and morphology it was intermediate between *C. lupus chanco* and a species represented by a domestic dog skull from a Chinese Neolithic site in Hemudu.

As the first synthesis of then known fossil records, Olsen's effort to make the connection between Chinese fossils and domestic dogs was certainly a worthy one. Olsen, however, realized that the co-occurrence of hominid and canid in the middle Pleistocene (around 500,000 years ago) was far too early to be associated with domestication events and thus cautioned that "although this association of hominids and wolves at this early period does not imply in any way either taming or early domestication, it does place both genera of animals in contemporary association that apparently continued until such time that these events did occur" (1985:42). His assumption of the continuity of hominids and canids in China is probably incorrect. Current knowledge suggests that the archaic *Homo erectus* likely had no direct relationship with *Homo sapiens*, who arrived much later in Eurasia directly from Africa (although an alternative multiregional hypothesis does exist that claims links between *H. erectus* and *H. sapiens*). On the canid side, our own research also indicates that the middle Pleistocene Zhoukoudian canid actually belongs to a species of its own, *Canis variabilis*, which is only distantly related to the modern wolves. Therefore, the Zhoukoudian materials probably have little relevance to dog domestication. This is not to say that China could not have been the center of origin for domestic dogs, but that the comparison of *C. variabilis* materials and domestic dogs is probably not appropriate.

At about the same time the Olsens (1977) proposed that domestic dogs had originated in China, two Israeli zooarchaeologists, Simon J. M. Davis and François R. Valla (1978), announced the discovery of a canid puppy buried with a human in Mallaha, an archaeological site near the old Huleh Lake in the upper Jordan Valley of Israel. The Mallaha site belonged to the Natufian culture, which is believed to have been one of the last hunter-gatherer groups, dating to approximately 12,000 to 10,000 years ago. The Natufian people lived in circular dwellings in what were perhaps the earliest permanently settled villages, foreshadowing later agricultural societies. However, it is the way a human was buried with a dog that really captured the archaeologists' imagination.

In the entrance to the dwelling, 25 cm below a large slab of limestone, which is often indicative of the presence of a burial, a human skeleton lies curled in a fetal position on its right side. The skeleton belongs to an individual of old age, judging by the wear on the teeth, but its sex is uncertain due to damage to the pelvic area. The left hand of this skeleton reaches across the head and is partly buried below the head. This hand is wrapped around the chest of a puppy, and the person's head is resting on top of the puppy. The intimacy of these two individuals is truly remarkable and argues strongly for a relationship that is more than a mere casual association.

Analysis of the teeth and of the fusion sequence of the long bones in the puppy skeleton indicates that it is of an individual around four to five months old. Adult canid materials were also discovered, including a lower jaw from the same dwelling and an isolated lower carnassial tooth (m1) from the terrace of Hayonim in western Galilee (the latter specimen is associated with a Natufian assemblage that has yielded a ^{14}C date of 11,920 B.P.). The size of these three specimens falls in the lower extremes of modern wolves in the Middle East, but is far smaller than late Pleistocene wolves (around 45,000 to 14,000 years ago), which tended to be much larger due to the colder climate at that time. However, the Natufian canids fall comfortably in the size range of living dogs and of fossil dogs found at archaeological sites, although they tend to be on the upper end of the range for domestic dogs. Analyses of other morphological features, such as the degree of crowding of the premolars, a character of domestic dogs due to the shortening of their rostrum, also place the Natufian canid in the transitional position between wolves and dogs. It thus seems reasonable to postulate that the Natufian canid belonged to an early dog species that has subsequently gotten smaller. Davis and Valla speculated that "the puppy . . . offers proof that an affectionate rather than gastronomic relationship existed between it and the buried person" and that it was "man-the-hunter" who had domesticated the wolf (1978:609).

This example brings us to another distinction of dogs: they and cats (which were domesticated later and also in the Near and Middle East) are the only carnivores among early domesticates. Such a distinction is important because carnivorans share the same food resources as humans. Large herbivores convert vegetation not usable by humans into consumable products such as meat and milk. It is difficult to imagine that dogs were first raised for food because they consume more humanly edible food than they produce. Therefore, for dogs to have been maintained with resources precious for humans themselves, the dogs must have served a vital function that was otherwise unobtainable. Davis and Valla (1978) were probably right that dogs did not serve as human food until much later, when agricultural surpluses could sustain more dogs (this is not to say that early humans did not eat dogs at all; they probably did opportunistically). Was it collaborative hunting or companionship that provided the compelling reason for a human to keep a dog?

Other discoveries of early domestic dogs have occasionally been reported in scholarly journals, but they are often difficult to verify due to the poor quality of the records. For example, German scientist Günter Nobis reported in 1979 a partial right jaw of a domestic dog from a Cro-Magnon site in Oberkassel, near Bonn. This jaw is dated to around 14,000 B.P. Only four teeth are left in the jaw (canine, fourth premolar, and first and second molars), and it appears that the second and third pre-molars were lost during life (the alveoli for the roots of these two teeth are no longer visible). If the remains' identity as a dog can be confirmed, this find would constitute one of the earliest records of domestic dogs, and Nobis speculated that the German dog may represent a domestication independent from those at the other sites in Eurasia. However, it is difficult to ascertain a dog identity for the Oberkassel speci-men, given its lack of skeletal regions that allow diagnosis, such as the forehead area and coronoid process, parts of the cranial anatomy that are important in the iden-tification of dogs. Consequently, specialists have not widely accepted this German record as evidence of early dog domestication.

Mikhail V. Sablin and Gennady A. Khlopachev (2002) have reported two nearly complete dog skulls from the upper Paleolithic site of Eliseevichi 1 in the Bryansk Region of the central Russian Plain. Radiocarbon dates on associated fauna (mostly mammoth, arctic fox, and reindeer) have yielded a range of 17,000 to 13,000 years B.P.—the upper range of the date indicating that this find would be the earliest re-cord for dogs. The two skulls are of large, northern wolf size. The main dog fea-tures on the Eliseevichi skulls are a short rostrum and a broad palate, characters that supposedly resemble those of the Siberian husky. Furthermore, on one of the skulls (MAE 447/5298), there is a large hole on the left side of the braincase, which Sablin and Khlopachev interpret to be a man-made break for access to the brain; that is, they postulate that its brain was consumed by humans. Intriguing as the new Russian materials may be, evidence for the argument that the Eliseevichi skulls belonged to domestic dogs appears to be ambiguous. The P4 on ZIN 23781(24) (the other skull from Eliseevichi) has a length of 27.3 mm, compared with the P3 length of 17.4 mm. The size proportion between P3 and P4 is that of a gray wolf rather than that of a dog, which would have a relatively reduced P3 to accommodate a shortened rostrum. However, one cannot rule out that the Eliseevichi canid was a very primi-tive, wolflike dog that was in a beginning phase of domestication—morphology at this stage of divergence is usually incapable of unambiguously resolving the differ-ence. DNA studies may help, but until some DNA can be extracted from the speci-mens, the identity of the Russian materials will likely remain uncertain.

Morphology of Domestic Dogs

Although most authors agree that some kind of a southern (Chinese or Arabian) wolf gave rise to domestic dogs because of the overwhelming morphological and genetic similarities between the wolf and the dog, these similarities become a confounding problem in distinguishing the remains of dogs from those of wolves at archaeological sites. As a result, the discoveries of supposed domestic dogs at the archaeological sites in China, Israel, and Europe have been disputed due to the absence of a clear criterion to distinguish dogs from wolves in fossil materials. For example, Stanley Olsen (1985) dismissed the Natufian materials as possibly an aberrant specimen of wolf, whereas Tamar Dayan (1994), an Israeli evolutionary biologist, marshaled further evidence that the Israeli burial materials were more likely be domestic dogs. Morphological characteristics, or the lack thereof, lie at the center of the issue.

We begin by looking at modern dogs in order to establish a firm basis of comparison. With a few exceptions, modern dogs are generally smaller than wolves. Along with smaller size, adult dogs also exhibit juvenile characters as though they have never grown up, including such features as short snouts, floppy ears, and large eyes. Biologists call these characters *paedomorphic features*, referring to the phenomenon that animals that have reached sexual maturity still retain juvenile features. Some evolutionary biologists have speculated that dogs with paedomorphic characters appeal to the human sense of affection. In other words, a "cuteness factor" seems to be at work in bridging the gap between dogs and humans. A paedomorphic dog may also be more submissive, a vital characteristic for a successful domestic relationship.

Skulls and jaws are the most commonly preserved elements at archaeological sites. Dogs tend to have a more domed forehead (easily seen in the profile view, which shows an obvious bulge above the orbit, or bony socket, of the eye) as well as a shortened rostrum and jaws. Associated with the short rostrum are more crowded premolars because there is less room in the jaws. To alleviate this crowding, the premolars are also reduced in size. In addition, some authors claim to have observed in dogs a narrower and more concave posterior border in profile view of the coronoid process (the upwardly protruding bony plate above the mandibular joint) of the lower jaw. The temporalis muscle inserts on this process. This muscle closes the jaw and is much more robustly developed in wolves. Again, its form in dogs is part of the paedomorphic form of the skull.

These morphological generalities work in identifying most modern domestic dogs. When it comes to distinguishing dogs from wolves in the remains found at archaeological sites, however, many of these characters break down as dogs become more wolflike as the archaeological records move closer to the origin of domestic dogs. At a certain point, it becomes almost impossible to ascertain the true identity of a doglike specimen at an archaeological site. At that point, a definitive human–dog association, as seen in the Natufian burial site, becomes more valuable.

A Russian Experiment on the Fox

Morphological characters of dogs are certainly important, particularly in archaeological context, but how dogs were modified behaviorally is equally insightful in gaining a comprehensive understanding of dog domestication. Behavioral considerations are especially relevant given the fact that one of the most critical factors in domestication is the successful modification of animal behavior toward humans. An elegant Russian experiment helped to make this point clear.

Dmitry K. Belyaev believed that the key factor that initiated all the genetic and morphologic changes of wolves into domestic dogs was behavior modification. In particular, the amenability to domestication, or tameability, was the most important factor in determining the success or failure of domestication. Belyaev and his comrades began working with foxes in a long-term experiment of more than 40 years (described in Trut 1999). The Russian geneticists selectively bred foxes for a single trait: tameness. After about 30 to 35 generations, a large percentage of the foxes showed the unmistaken signs of domestic animals: they were docile and eager to please. Remarkably, these behavioral selections also brought about several physical changes. The foxes began to have floppy ears, short and rolled tails, short legs, lighter coat colors, and malocclusion of the teeth (the upper jaw shorter than the lower jaw, as seen in some bulldogs)—characters common in domestic dogs, as Charles Darwin noted long ago.

Belyaev explained his experiment in terms of selection for tameability and its consequence for developmental processes. As noted, many of the physical and behavioral characters in dogs may be seen in light of the paedomorphic process—that is, the retention of juvenile features in adults. After all, to be tame is to shut down the fear response, a behavior that serves wild species well. In a sense, domestic dogs never grow up and are stuck in juvenile stages both behaviorally and morphologically.

Genetic History of Domestic Dogs

Whereas archaeologists are rarely able to find more than a few fragmentary fossils to hint at the records of early dogs, molecular biologists have increasingly turned to DNA for information about the history of dogs. Changes in DNA over the generations are often preserved in the vast and rich sources of the dog genome. In December 2005, a complete map of the DNA code of a female boxer was published by a group led by Kerstin Lindblad-Toh. The dog genome, as it is commonly known for this kind of project, was the first to be sequenced among the 271 species of carnivorans. In fact, the complete genome of only a few organisms has been sequenced;

among mammals, the dog is only the fifth species to be sequenced (the first four are the human, the chimpanzee, the mouse, and the rat). The selection of the dog to reveal its complete genetic code thus highlights the dog's special status in human society and its importance in medical and scientific research.

Among the results of the dog genome study is the ability to identify a small collection of rapidly evolving sequences in various parts of the dog genome that can be used to trace the phylogenetic relationships among living canids. A selection of about 11,100 base pairs of DNA for 30 of the 34 living canid species has permitted the reconstruction of the phylogenetic tree of the family Canidae. Not surprisingly, domestic dogs and wolves are closely related, as morphologists and molecular biologists have long suspected.

Molecular biology has also contributed to a recent controversy over the time and place of origin of the first domestic dogs. Carles Vilà and his colleagues proposed in 1997 that multiple domestication events happened as early as 100,000 years ago. Such an estimate is based on the concept of the molecular clock, which assumes a steady rate of change in molecular composition in a certain region of the DNA. By using one or more existing fossil records as a minimum starting point for the origin of a particular species, a molecular biologist can estimate the rate of base-pair changes in a given amount of time. For example, Vilà and his colleagues used 1 million years as the minimum time of divergence between wolf and coyote. Thus a small amount of molecular difference in the mitochondrial DNA sequence between wolf and dog can be translated to a time period of 100,000 years since the divergence of these canines.

This estimate places dog domestication far earlier than any of the existing fossil evidence indicates. Vilà and his group suggested that early domestic dogs were far too wolflike to be recognizable in the fossil record and that it was only after about 15,000 years ago that artificial selection achieved enough morphological difference for us to be able to recognize dogs in the archaeological records.

In contrast to the study by Vilà and his colleagues (1997), Peter Savolainen and his co-workers (2002) contend that the first domestication of dogs was done in East Asia about 15,000 years ago. Savolainen analyzed a large sample of 654 dog breeds, each with 582 base pairs of mitochondrial DNA, and found a larger genetic variation in East Asia than in other regions, which suggests an East Asian origin for the domestic dog. Because mitochondrial DNA is inherited from the maternal side, Savolainen's group could trace modern dogs to five female wolf lines. Of these lines, a lineage ("Clade A") that includes three closely linked positions in Chinese and Mongolian wolf genes has the highest genetic diversity and thus was postulated to contain the ancestral population of domestic dogs.

Dogs of the New World

Despite these controversies, a general consensus exists that dogs were domesticated prior to 15,000 B.P. and that domestic dogs originated in Eurasia. But what happened to dogs elsewhere? In particular, what is the relationship between New World dogs and their Old World counterparts? The dog was the only domesticated species that was present across Eurasia and the Americas before the arrival of Columbus in the fifteenth century. Wolves, in contrast, had a Holarctic (the northern continents of Europe, Asia, and North America) distribution and thus had every opportunity to be associated with humans when they crossed the Beringian land bridge. As noted, fossil evidence indicates that dogs were domesticated around 14,000 to 12,000 years ago, a time period that either closely coincides with the first arrival of humans in the New World, about 15,000 years ago, or even slightly lags behind the human crossing of Beringia. If this timing is taken literally, then dogs could not have been with the first humans to come to the Americas. The questions we have to ask therefore are: Was there any connection between the New World dogs and the Old World dogs, or were the New World dogs independently domesticated by the Native Americans?

To answer these questions, Jennifer A. Leonard and her colleagues (2002) devised another genetic study of dogs. Modern New World dogs cannot be used in such a study because their ancestors were likely to have been interbred with dogs brought by European colonists, and their genetic makeup would not reveal their ancestral conditions. Leonard and her colleagues, therefore, extracted DNA from the bones of 37 dog specimens recovered from pre-Columbian archaeological sites in Mexico, Peru, and Bolivia, which ensured that no interbreeding with European dogs had occurred. Analysis of this ancient DNA revealed that native American dogs are closely related to Eurasian ancestral lineages, suggesting that dogs accompanied late Pleistocene humans when they crossed the Beringian landmass. If so, dogs must have played a remarkable role in the early occupation of the Americas, a case in which a domestic animal assisted humans to settle in a new world.

There is also circumstantial fossil evidence that humans brought their dogs along when they first crossed Beringia to arrive in North America around 15,000 years ago. Ted Galusha, the late collector and curator of the Frick Laboratory at the American Museum of Natural History in New York, noted that several skulls from the late Pleistocene deposits near Fairbanks, Alaska, were extremely short-faced for wild wolves and approached the facial proportions of modern Eskimo dogs. Galusha did not publish his results, which he turned over to Stanley Olsen for further study. Olsen was convinced that these fossil skulls belonged to the forerunners of the Eskimo dogs. However, true wolves were also undoubtedly present in the same deposits, making it difficult to resolve the question of whether the short-faced individuals were part of the range of variation within the wolf population. The genetic study of 11 samples from Alaska by Leonard and her colleagues (2002) confirms the individuals' identity as dogs.

Australian Dingo

The Australian dingo (*Canis lupus dingo*) is another interesting example of humans' first domestic animal braving a new world together with its masters. When the first Europeans arrived in Australia in the eighteenth century, the dingo was the only large placental mammal there, besides the Australian Aborigines, among a diverse group of marsupials. The name "dingo" comes from the Eora Aboriginal tribe of southeastern Australia. In external morphology, dingoes, with their tan coat color, closely resemble domestic dogs, particularly those from southern Asia. In their behavior, however, they tend to be more independent than dogs, and they roam wild over Australia. Although occasionally used for pets or hunting, dingoes generally resist full domestication by modern humans.

Almost from the beginning of the Europeans' arrival, there was debate about the dingo's possible origin as a feral dog and its possible relationship with southern Asian dogs. Did it arrive in Australia as a semidomestic animal and then secondarily become feral, or did it come on its own as a wild canid? Because there was substantial open water of at least 50 km between islands along the Southeast Asian archipelago, even during maximum glaciation when the sea level was at its lowest, could a wild canid have swum across the sea without human aid? Or were dingoes carried on boats by Southeast Asian fisherman and sea cucumber harvesters? Rodents and bats are the only placental mammals to have made it to Australia on their own, presumably by island hopping, rafting, and flying from Southeast Asian islands. No large mammal has made the journey, and traveling by boat with humans seems to be the most likely scenario for the dingoes.

Dingoes closely resemble southern Asian dogs and wolves morphologically. In archaeological records, the earliest definite dingo remains date to approximately 3,500 years ago. Peter Savolainen and his colleagues (2004) again applied molecular techniques to answer questions about the connection between New World and Old World dogs. Their analysis of mitochondrial DNA from 211 Australian dingoes, wolves, and domestic dogs suggests that dingoes largely have the genetic makeup of domestic dogs. They share a particular gene type called A29 with dogs from East Asia and arctic America. By calculating the amount of genetic divergence, Savolainen's group concluded that dingoes arrived in Australia around 5,000 years ago in a single migration event, a date roughly consistent with the archaeological dates. If these dates from the archaeological and the molecular evidence are anywhere close to the true time of the dingoes' arrival, then the dingoes must have come by boat in a much later wave of human dispersal than that of the first settlers of Australia, about 50,000 years ago.

Despite these studies, questions remain about how dingoes arrived in Australia. Did they become feral because they were not fully domesticated in the first place? Or did they find easy prey in the form of marsupials and thus have little use for humans after landing in Australia? On an island continent without other placental predators, dingoes probably faced relatively light competition and found it easy to stay wild. Whatever the actual scenario, dingoes are a unique example of a carnivoran that invaded a new continent by an unconventional means (boats), through either natural association with or domestication by humans, and ultimately became the top dog in the community.

Why Were Dogs Domesticated First?

As noted, dogs were the first animal to be fully domesticated by human hunter-gatherers, and, as such, they played a unique role in human history, instilling the idea that animals could be harnessed for human purposes. As the first and presumably the only domestic animal at the time, dogs accompanied humans to the Americas and, in a sense, used human technology (the boat) to expand into Australia. They are thus pivotal in the evolution of human societies, and, to a certain extent, their importance is on a par with that of such key inventions as stone tools, bronze, and agriculture. One is thus tempted to ask the question: Why were dogs domesticated first? Why did it take a highly social carnivore to give humans the idea of domestication? Did the mammal have to be intelligent to be able to coexist with us? Or had dogs reached a level of sociability similar to that of humans? The beginning of sedentary human communities is associated with the origin of agriculture. Perhaps dogs were the only domestic animal that fit the hunter-gatherer lifestyle. Perhaps humans observed and learned from wolves' pack-hunting techniques? Could it be, as Stanley Olsen (1985) speculated, that humans as pack hunters themselves were brought close to the wolves because of their shared tactics in obtaining prey?

Or perhaps hunting had nothing to do with the early contact of humans and wolves. Olsen was probably the first to put this idea on the table: "It is possible that hungry wolves were enticed (not necessarily by human design) to come closer to a campfire, where meat was being cooked and the refuse discarded in the immediate vicinity of the camp. Perhaps wolves that had attached themselves loosely to human habitation areas would consider such camps as their home territory, and the warning growls toward intruders would also warn the human inhabitants of the approach of such outsiders" (1985:18).

The conventional notion of domestication often contains an element of human design. The idea that early humans intentionally selected certain desirable traits or eliminated undesirable traits, thereby creating a breed that suited certain human needs, is rooted in the modern practice of artificial selection in agriculture and animal husbandry. Although such societal intentions are part of decision making when it comes to modern practices of domestication, human motivations in historical context are difficult, if not impossible, to fathom and cannot be demonstrated empirically by the archaeological records. To circumvent this difficulty, Darcy F. Morey (1994) proposed a different approach to the question of the initial domestication of dogs. He suggested that it is not necessary to presume human intentions to make sense of early domestication.

Morey indicated that an evolutionary perspective would do away with anthropocentric approaches. He argued that focusing on the human role in domestication ignores the evolutionary stakes for the participating animals. The success of the dog–human association in almost every conceivable environment on every continent suggests that dogs have profited well from this arrangement, particularly in contrast to the dwindling wolf populations worldwide. Perhaps the ancestral dogs (or wolves) found such an association advantageous and thus willingly participated in an experiment that eventually benefited both parties. Such a view does not have to invoke the conscious act of humans to tame a wild carnivore and thus is free from the intractable and messy arguments about human intentions.

Morey (1994) suggested that late Pleistocene hunter-gatherers (from around 12,000 years ago) were in regular contact with wolves because both hunted many of the same prey. Some wolf pups were then probably incorporated into a human social setting, and some adopted pups survived to adulthood. These wolves must have learned that subordinate status to dominant humans was the condition to be tolerated in human society and that being able to solicit food from people was also a valuable skill. The diet of wolves that lived around human settlements must have changed as well, from predominantly meat to a variety of meat and plants, some of it garnered or scavenged from humans' scraps. In such a domestic setting, smaller body size is also a favorable trait. Small size and submissive behavior can easily be achieved by paedomorphosis during individual development.

Behavioral biologists Raymond Coppinger and Lorna Coppinger (2001) have taken Morey's (1994) model of the dog's self-domestication a step further. They envision the following scenario for dog domestication. First, agriculture created human settlements, a way of living that contrasted with the nomadic hunter-gatherer lifestyle. In every human village, there were discarded products such as bones, carcasses, grains, fruits, as well as human waste. The Coppingers argue that this human dump site became the first niche for some wolves. These wolves would frequent the garbage dump to gain access to the new food source. Those wolves that were less frightened by humans tended to be more successful in making a living this way because they would waste less energy evading humans when they saw them approach. Such wolves by definition were more tame, thereby leading to the early association of wolves and humans, which ultimately led to the domestication of dogs.

One of the important distinctions of the Coppingers' (2001) model is that it does away with wolves' sociality in the equation. They argue that sociality has nothing to do with domestication. In fact, they suggest that the incipient dogs subsisted on low-quality food and must have had smaller brains (along with a smaller body size, smaller heads, and smaller teeth) than wild wolves, which had to coordinate the pack (chapter 5). Contrary to the popular belief that dogs possess a high intelligence, the Coppingers point out that dogs not only have a smaller brain than wolves, but also are less intelligent.

Why Were Canids Domesticated?

Although the garbage-dump model is an attractive alternative, perhaps the biggest difficulty with Raymond and Lorna Coppinger's (2001) hypothesis is that once the sociality of wolves is deemed irrelevant, the domestication of dogs is no longer a unique association of two social species: *Homo sapiens* and *Canis lupus*. Other hypocarnivorous carnivorans are also adept at feeding on human waste dumps. Raccoons and bears are two well-known examples, although the former were not in contact with humans until the Paleo-Indians arrived in North America, at which time humans already had dogs as their companions. Bears, though, were around Mesolithic hunter-gatherers, but were never domesticated. But bears are not social predators and may not be easily domesticable.

Before we address the question of why canids first, let us broaden the perspective somewhat to consider the spectrum of early domestic animals. With the exception of cats, which had a relatively minor economic role (catching mice) in early human societies, dogs stood alone as the only carnivore among 14 successfully domesticated large herbivores. If we get past the anthropocentric notion of domestication as a human invention, which Darcy Morey (1994) and the Coppingers (2001) suggest we do, then humans might have been educated about the possibility of living with other animals by association with certain species. However, early humans were hunter-gatherers, and most of the large herbivores were probably their prey. It is therefore difficult to envision that any of the prey species were not especially guarded when encountering humans. Furthermore, there is little that would entice these herbivores to come close to human settlements; the same human waste dumps that were attractive to the wolves as a supplement to their diets were hardly worth visiting from the herbivores' point of view.

In contrast, carnivorans as predators are less afraid of approaching other species; they have to do it routinely in order to catch their prey. This is not to say that early humans and wolves did not have an antagonistic relationship; all top predators occasionally confront other predators either to loot the other's spoils or to attack the other. But it is perhaps far easier for a carnivoran to approach another predator, such as a human, than it is for an ungulate herbivore to approach a predator. Therefore, it may have been inevitable that the first domestic animal came from a member of the Carnivora.

Various families of carnivorans were part of the mammalian fauna surrounding early *Homo sapiens*. We can probably quickly eliminate Mustelidae (weasels, badgers, and otters), Viverridae (civets), Herpestidae (mongooses), and Procyonidae (raccoons and ringtail "cats") as families of small carnivorans that could not play the same role as dogs in human society (the New World Procyonidae were not even in contact with humans until the Paleo-Indians expanded into the Americas). That elimination leaves four families of large carnivorans to consider: Felidae (cats, lions, and tigers), Ursidae (bears and giant panda), Hyaenidae (hyenas and aardwolves), and Canidae. Although some felids can be highly social predators, they are hypercarnivores (chapter 4) that have essentially a pure meat diet. In other words, human dump sites were probably not appealing to them and, in any case, probably could not have provided them with enough food for subsistence (cats obviously were an exception because they are small enough not to pose a burden on humans). Ursids, as a generalized omnivore, would also have found garbage dumps attractive, but, as noted, bears do not have the right temperament to be readily tamed. Even if they could be tamed, they lack the wolves' running abilities and hunting skills and thus would not have contributed enough for a long-term relationship with humans. Finally, hyaenids are close enough to canids in their dietary requirements and social behavior to have frequented human dump sites. Hyaenids, however, appear to lack the requisite temperament to be domesticated.

Thus Canidae is the only group of carnivorans that possesses the right qualities to be domesticated. Canids are not too large and thus can be dominated by humans, and they are mesocarnivorous enough to be able to handle a variety of foods. Such a dietary preference probably predisposed canids to scavenge human garbage dumps. A final critical factor is dogs' temperament. Among all domestic animals, dogs are among the most submissive to humans. Even if we discount the longer history of dog domestication, dogs seem to be more predisposed toward humans than other domestic animals. Cats, for example, have almost an equally long history of domestication (approximately 9,000 years), but as all cat owners know, cats are far more independent and less solicitous toward humans—millennia of artificial selection has not brought about a cat that is as submissive to humans as dogs are. This submissive attitude toward humans probably also facilitated the training for other functions once dogs fully integrated themselves into human societies: as companions, hunting partners, sentinels, guards, and so on.

In a study of social cognition in dogs, Brian Hare and his colleagues (2002) suggest that dogs possess a high level of social-communicative skills with humans. Their study measured the ability of dogs to follow visual, gestural, and other cues to accomplish tasks. Dogs in such experiments outperformed chimpanzees and wolves in comprehending human communicative signals. Hare and his colleagues concluded that dogs' ability to use human social cues was developed in the process of domestication. We caution, however, that many studies demonstrating special abilities by domestic animals are probably testing mental capacities acquired during long processes of domestication and that these skills are not likely to have been present during initial domestication.

Regardless of dogs' initial capability, a combination of morphological and behavioral characteristics seems to have made them uniquely qualified to be incorporated into human society. In a sense, it was almost inevitable that dogs were the first animals to have a mutually beneficial relationship with humans.

The Dizzying Array of Dog Breeds

With the possible exception of the goldfish, with its often bizarre morphology artificially bred from a species of carp, domestic dogs are perhaps the most differentiated from their wild ancestors. The wonderful variety of sizes, shapes, and coat colors they have would make an unsuspecting taxonomist place them in separate families if he or she did not know that these traits are artificially selected. By contrast, domestic cats are far more uniform in their morphology, and coat color is the main source of variation. Such a tremendous amount of variation within a single subspecies of wolves is almost inconceivable in nature. (These various traits have no survival value and would have been quickly eliminated if they were to occur naturally.) Nonetheless, these variations seem to indicate a vast reservoir of genetic variability not seen in other mammal species.

Why are dogs so variable? One clue may lie in why cats are so invariable. Jill A. Holliday and Scott J. Steppan (2004) attempted to address this question by comparing the morphological diversity of fossil cats and dogs. Cats are extreme hypercarnivores, and their entire muscle-skeletal-dental system seems to be perfected for a single purpose—to catch prey. With the exception of the saber-toothed cats, much of the history of cat evolution consists of species that resemble one another closely—so closely, in fact, that the fossils of different extinct cats are often difficult to distinguish from one another. It appears that cats' cranial and dental morphology has been fine-tuned to the point that changes to such a body plan would perhaps be detrimental to their survival, so natural selection has tended to eliminate these changes. Holliday and Steppan thus ask the question: Does increasing hypercarnivory limit morphological variation? Framed in scientific terms, the question becomes: Are increasing degrees of specialization correlated with decreasing phenotypic variation?

The answer to this question is "yes": specialization to hypercarnivory has a strong effect on subsequent morphological diversity, particularly regarding the jaws and dentition. In layman's terms, this means that cats tend to be locked in their morphological system or, alternatively, that there has been a strong selective pressure to eliminate variations from a nearly perfect system. Canids, however, are more flexible in their dental and cranial morphology because primitively they are mesocarnivores. Therefore, evolutionarily speaking, canids appear to have more built-in flexibility than felids have.

However, we need to be cautious against too literal an application of this discovery. For one thing, Holliday and Steppan's (2004) measurements are of skeletal and dental areas that are often evolutionarily the most variable. These areas do not necessarily change that much among dogs. For example, the relative proportion of the shearing and grinding part of the dentition, an indication of the degree of hypercarnivory (chapter 4), is not nearly as variable as the proportions of other parts of the dog skeleton. Instead, much of the skeletal variation in dogs is related to the timing of development. In other words, paedomorphosis probably played a larger role in the great morphological variations in the body size and cranial proportions of dogs.

APPENDIX 1
CANID SPECIES
AND CLASSIFICATION

THIS LIST INCLUDES SPECIES OF CANIDS, their known geologic age, and their known geographic range. For living canids, we follow Wozencraft (1993); Tedford, Taylor, and Wang (1995); and Wang, Tedford, et al. (2004b). For fossil canids, we follow Wang (1994) for Hesperocyoninae; Wang, Tedford, and Taylor (1999) for Borophaginae; and Berta (1981, 1987, 1988) for South American Caninae. A forthcoming monograph on North American fossil Caninae by Tedford, Wang, and Taylor describes a number of new species; these are variously listed as "sp. A," "sp. B," and so on, in order to avoid nomenclatural confusions before they are officially published. There is currently no comprehensive synthesis on Eurasian and African canids, and our list is a compilation from various literatures, and we did not include many of the poorly documented species or species from questionable records. The general sequence of the genera and species follows their approximate position in phylogeny; that is, primitive taxa are listed first and followed by more advanced taxa. The abbreviations are: Af, Africa; As, Asia; E, Early; Eoc, Eocene; Eu, Europe; L, Late; M, Middle; Mio, Miocene; NAm, North America; Oligo, Oligocene; Pleist, Pleistocene; Plio, Pliocene; R, Recent; SAm, South America.

SCIENTIFIC NAMES, COMMON NAMES (FOR LIVING SPECIES)	AGE	DISTRIBUTION
Subfamily Hesperocyoninae Martin, 1989		
Prohesperocyon wilsoni (Gustafson, 1986)	L Eo	NAm
Hesperocyon gregarius (Cope, 1873)	L Eo–M Oligo	NAm
Hesperocyon coloradensis Wang, 1994	E Oligo	NAm
Mesocyon coryphaeus (Cope, 1879)	L Oligo–E Mio	NAm
Mesocyon temnodon (Wortman and Matthew, 1899)	E–L Oligo	NAm
Mesocyon brachyops Merriam, 1906	L Oligo–E Mio	NAm
Cynodesmus thooides Scott, 1893	E–L Oligo	NAm
Cynodesmus martini Wang, 1994	M Oligo	NAm
Sunkahetanka geringensis (Barbour and Schultz, 1935)	M Oligo	NAm
Philotrox condoni Merriam, 1906	M Oligo	NAm

Enhydrocyon stenocephalus Cope, 1879	M–L Oligo	NAm
Enhydrocyon pahinsintewakpa (Macdonald, 1963)	M–L Oligo	NAm
Enhydrocyon crassidens Matthew, 1907	E Mio	NAm
Enhydrocyon basilatus Cope, 1879	L Oligo–E Mio	NAm
Osbornodon renjiei Wang, 1994	E Oligo	NAm
Osbornodon sesnoni (Macdonald, 1967)	E Oligo	NAm
Osbornodon wangi Hayes, 2000	E Mio	NAm
Osbornodon scitulus (Hay, 1924)	E Mio	NAm
Osbornodon iamonensis (Sellards, 1916)	E Mio	NAm
Osbornodon brachypus (Cope, 1881)	E Mio	NAm
Osbornodon fricki Wang, 1994	E–M Mio	NAm
Paraenhydrocyon josephi (Cope, 1881)	E Oligo–E Mio	NAm
Paraenhydrocyon sp.	M Mio	As
Paraenhydrocyon robustus (Matthew, 1907)	E Mio	NAm
Paraenhydrocyon wallovianus (Cope, 1881)	L Oligo–E Mio	NAm
Caedocyon tedfordi Wang, 1994	L Oligo	NAm
Ectopocynus antiquus Wang, 1994	E–L Oligo	NAm
Ectopocynus intermedius Wang, 1994	M–L Oligo	NAm
Ectopocynus simplicidens Wang, 1994	E Mio	NAm

Subfamily Borophaginae Simpson, 1945

BASAL BOROPHAGINAE

Archaeocyon pavidus (Stock, 1933)	M Oligo	NAm
Archaeocyon leptodus (Schlaikjer, 1935)	E–L Oligo	NAm
Archaeocyon falkenbachi Wang, Tedford, and Taylor, 1999	L Oligo	NAm
Oxetocyon cuspidatus Green, 1954	E Oligo	NAm
Otarocyon macdonaldi Wang, Tedford, and Taylor, 1999	E Oligo	NAm
Otarocyon cooki (Macdonald, 1963)	M–L Oligo	NAm
Rhizocyon oregonensis (Merriam, 1906)	M Oligo	NAm

TRIBE PHLAOCYONINI WANG, TEDFORD, AND TAYLOR, 1999

Cynarctoides lemur (Cope, 1879)	M–L Oligo	NAm
Cynarctoides roii (Macdonald, 1963)	M Oligo–E Mio	NAm
Cynarctoides harlowi (Loomis, 1932)	E Mio	NAm
Cynarctoides luskensis Wang, Tedford, and Taylor, 1999	E Mio	NAm
Cynarctoides gawnae Wang, Tedford, and Taylor 1999	E Mio	NAm
Cynarctoides whistleri Wang and Tedford, 2008	E Mio	NAm
Cynarctoides acridens (Barbour and Cook, 1914)	E–M Mio	NAm
Cynarctoides emryi Wang, Tedford, and Taylor, 1999	E Mio	NAm
Phlaocyon minor (Matthew, 1907)	M Oligo–E Mio	NAm
Phlaocyon latidens (Cope, 1881)	M Oligo	NAm
Phlaocyon annectens (Peterson, 1907)	E Mio	NAm
Phlaocyon taylori Hayes, 2000	E Mio	NAm

Phlaocyon achoros (Frailey, 1979)	L Oligo	NAm
Phlaocyon multicuspus (Romer and Sutton, 1927)	E Mio	NAm
Phlaocyon marslandensis McGrew, 1941	E Mio	NAm
Phlaocyon leucosteus Matthew, 1899	E–M Mio	NAm
Phlaocyon yatkolai Wang, Tedford, and Taylor, 1999	E Mio	NAm
Phlaocyon mariae Wang, Tedford, and Taylor, 1999	E Mio	NAm

TRIBE BOROPHAGINI WANG, TEDFORD, AND TAYLOR, 1999

BASAL BOROPHAGINI

Cormocyon haydeni Wang, Tedford, and Taylor, 1999	L Oligo–E Mio	NAm
Cormocyon copei Wang and Tedford, 1992	L Oligo–E Mio	NAm
Desmocyon thomsoni (Matthew, 1907)	E Mio	NAm
Desmocyon matthewi Wang, Tedford, and Taylor, 1999	E Mio	NAm

▪ SUBTRIBE CYNARCTINA MCGREW, 1937

Paracynarctus kelloggi (Merriam, 1911)	E–M Mio	NAm
Paracynarctus sinclairi Wang, Tedford, and Taylor, 1999	M Mio	NAm
Cynarctus galushai Wang, Tedford, and Taylor, 1999	M Mio	NAm
Cynarctus marylandica (Berry, 1938)	M Mio	NAm
Cynarctus saxatilis Matthew, 1902	M Mio	NAm
Cynarctus voorhiesi Wang, Tedford, and Taylor, 1999	M Mio	NAm
Cynarctus crucidens Barbour and Cook, 1914	M–L Mio	NAm
Metatomarctus canavus (Simpson, 1932)	E Mio	NAm
Euoplocyon spissidens (White, 1947)	E Mio	NAm
Euoplocyon brachygnathus (Douglass, 1903)	M Mio	NAm
Psalidocyon marianae Wang, Tedford, and Taylor, 1999	M Mio	NAm
Microtomarctus conferta (Matthew, 1918)	E–M Mio	NAm
Protomarctus optatus (Matthew, 1924)	E–M Mio	NAm
Tephrocyon rurestris (Condon, 1896)	M Mio	NAm

▪ SUBTRIBE AELURODONTINA WANG, TEDFORD, AND TAYLOR, 1999

Tomarctus hippophaga (Matthew and Cook, 1909)	M Mio	NAm
Tomarctus brevirostris Cope, 1873	M Mio	NAm
Aelurodon asthenostylus (Henshaw, 1942)	M Mio	NAm
Aelurodon montanensis Wang, Wideman, Nichols, and Hanneman, 2004	M Mio	NAm
Aelurodon mcgrewi Wang, Tedford, and Taylor, 1999	M Mio	NAm
Aelurodon stirtoni (Webb, 1969)	M Mio	NAm
Aelurodon ferox Leidy, 1858	M Mio	NAm
Aelurodon taxoides (Hatcher, 1893)	L Mio	NAm

▪ SUBTRIBE BOROPHAGINA WANG, TEDFORD, AND TAYLOR, 1999

Paratomarctus temerarius (Leidy, 1858)	M Mio	NAm
Paratomarctus euthos (McGrew, 1935)	M–L Mio	NAm

Carpocyon compressus (Cope, 1890)	M Mio	NAm
Carpocyon webbi Wang, Tedford, and Taylor, 1999	M–L Mio	NAm
Carpocyon robustus (Green, 1948)	M–L Mio	NAm
Carpocyon limosus Webb, 1969	L Mio	NAm
Protepicyon raki Wang, Tedford, and Taylor, 1999	M Mio	NAm
Epicyon aelurodontoides Wang, Tedford, and Taylor, 1999	L Mio	NAm
Epicyon saevus (Leidy, 1858)	L Mio	NAm
Epicyon haydeni Leidy, 1858	L Mio	NAm
Borophagus littoralis VanderHoof, 1931	L Mio	NAm
Borophagus pugnator (Cook, 1922)	L Mio	NAm
Borophagus orc (Webb, 1969)	L Mio	NAm
Borophagus parvus Wang, Tedford, and Taylor, 1999	L Mio	NAm
Borophagus secundus (Matthew and Cook, 1909)	L Mio	NAm
Borophagus hilli (Johnston, 1939)	L Mio	NAm
Borophagus dudleyi (White, 1941)	L Mio	NAm
Borophagus diversidens Cope, 1892	Plio	NAm

Subfamily Caninae Fischer de Waldheim, 1817

BASAL CANINAE

Leptocyon mollis (Merriam, 1906)	M Oligo	NAm
Leptocyon sp. A	M Oligo	NAm
Leptocyon vulpinus (Matthew, 1907)	E Mio	NAm
Leptocyon delicatus (Loomis, 1932)	L Oligo	NAm
Leptocyon gregorii (Matthew, 1907)	L Oligo	NAm
Leptocyon sp. B	E–L Mio	NAm
Leptocyon vafer (Leidy, 1858)	M–L Mio	NAm
Leptocyon sp. C	L Mio	NAm
Leptocyon sp. D	L Mio	NAm

TRIBE VULPINI, HEMPRICH AND EHRENBERG, 1932

Vulpes lagopus (Linnaeus, 1758), arctic fox	L Pleist–R	As, Eu, NAm
Vulpes stenognathus Savage, 1942	L Mio	NAm
Vulpes sp. A	L Mio	NAm
Vulpes riffautae Bonis et al., 2007	L Mio	Af
Vulpes alopecoides (Major, 1873)	L Plio	Eu
Vulpes beihaiensis Qiu and Tedford, 1990	E Plio	As
Vulpes praeglacialis Kormos, 1932	E–M Pleist	Eu
Vulpes chikushanensis Young, 1930	L Plio	As
Vulpes praecorsac Kormos, 1932	M Pleist	Eu
Vulpes angustidens Thenius, 1954	E Pleist	Eu
Vulpes galaticus Ginsburg, 1998	Plio	Eu
Vulpes bengalensis (Shaw, 1800), Bengal fox	R	As
Vulpes cana Blanford, 1877, Blanford's fox	R	As

Vulpes chama (Smith, 1833), Cape fox	R	Af
Vulpes corsac (Linnaeus, 1768), Corsac fox	R	As
Vulpes ferrilata (Hodgson, 1842), Tibetan fox	R	As
Vulpes macrotis Merriam, 1888, kit fox	R	NAm
Vulpes pallida (Cretzschmar, 1826), pale fox	R	Af
Vulpes rueppelli (Schinz, 1825), Rüppell's fox	R	Af, As
Vulpes velox (Say, 1823), swift fox	Plio–R	NAm
Vulpes vulpes (Linnaeus, 1758), red fox	Pleist–R	As, Eu, Af, NAm
Vulpes zerda (Zimmerman, 1780), fennec fox	L Pleist–R	Af
Metalopex sp. A	L Mio	NAm
Metalopex merriami Tedford and Wang, 2008	L Mio	NAm
Metalopex sp B.	L Mio	NAm
Urocyon sp. A	L Mio	NAm
Urocyon progressus Stevens, 1965	E Plio	NAm
Urocyon sp. B	Plio	NAm
Urocyon sp. C	Pleist	NAm
Urocyon minicephalus Martin, 1974	Pleist	NAm
Urocyon cinereoargenteus (Schreber, 1775), gray fox	Pleist–R	NAm
Urocyon littoralis (Baird, 1858), island gray fox	L Pleist–R	NAm
Prototocyon curvipalatus (Bose, 1879)	Pleist	As
Prototocyon recki Pohle, 1928	E Pleist	Af
Otocyon megalotis (Desmarest, 1822), bat-eared fox	L Pleist–R	Af

TRIBE CANINI FISCHER DE WALDHEIM, 1817

■ *SUBTRIBE CERDOCYONINA TEDFORD, WANG, AND TAYLOR, IN PRESS*

Atelocynus microtis (Sclater, 1882), short-eared dog	R	SAm
Cerdocyon sp. A	E Plio	NAm
Cerdocyon? avius Torres and Ferrusquia, 1981	E Plio	Am
Cerdocyon thous Hamilton Smith, 1839, crab-eating fox	Pleist–R	SAm
Chrysocyon sp. A	E Plio	NAm
Chrysocyon brachyurus (Illiger, 1815), maned wolf	Pleist–R	SAm
Dusicyon australis (Kerr, 1792), Falkland Island fox	L Plio–R	SAm
Nyctereutes donnezani (Depérer, 1890)	L Mio–Plio	Eu
Nyctereutes megamastoides (Pomel, 1842)	M–L Plio	Eu
Nyctereutes tingi Tedford and Qiu, 1991	Plio	As
Nyctereutes sinensis (Schlosser, 1903)	Plio	As
Nyctereutes abdeslami Geraads, 1997	L Plio	Af
Nyctereutes terblanchei Ficcarelli et al., 1984	Plio–Pleist	Af
Nyctereutes procyonoides (Gray, 1834), raccoon dog	Plio–R	As, Eu, Af
Pseudalopex culpaeus (Molina, 1782), culpeo	Pleist–R	SAm
Pseudalopex griseus (Gray, 1837), SAm gray fox	R	SAm
Pseudalopex gymnocercus (Fischer, 1814), pampas fox	R	SAm
Pseudalopex sechurae (Thomas, 1900), Sechuran fox	R	SAm
Pseudalopex vetulus (Lund, 1842), hoary fox	R	SAm
Speothos pacivorus Lund, 1839	L Pleist	SAm

Speothos venaticus (Lund, 1842), bush dog	L Pleist–R	SAm
Theriodictis? sp. A	Plio–Pleist	NAm
Theriodictis tarijensis (Ameghino, 1902)	Pleist	SAm
Theriodictis platensis Mercerat, 1891	Pleist	SAm
Protocyon troglodytes (Lund, 1838)	L Plio	SAm
Protocyon scagliarum Kraglievich, 1952	L Plio	SAm

■ *SUBTRIBE CANINA FISHER DE WALDHEIM, 1817*

Eucyon sp. A	L Mio	NAm
Eucyon davisi (Merriam, 1911)	L Mio	NAm, As
Eucyon zhoui Tedford and Qiu, 1996	E–M Plio	As
Eucyon intrepidus Morales, Pickford, and Soria, 2005	L Mio	Af
Eucyon monticinensis (Rook, 1992)	L Mio	Eu
Eucyon adoxus (Martin, 1973)	E Plio	Eu
Eucyon odessanus (Odintzov, 1967)	E Plio	Eu
Nurocyon chonokhariensis Sotnikova, 2006	Plio	As
"Canis" cipio Crusafont-Pairó, 1950	L Mio	Eu
Canis ferox Miller and Carranza-Castañeda, 1998	L Mio–E Plio	NAm
Canis lepophagus Johnston, 1938	Plio	NAm
Canis sp. A	L Plio	NAm
Canis sp. B	Pleist	NAm
Canis cedazoensis Mooser and Dalquest, 1975	L Pleist	NAm
Canis edwardii Gazin, 1942	L Plio–Pleist	NAm
Canis latrans Say, 1823, coyote	Pleist–R	NAm
Canis rufus Audubon and Bachman, 1851, red wolf	L Pleist–R	NAm
Canis armbrusteri Gidley, 1913	Pleist	NAm
Canis dirus Leidy, 1858	L Pleist	NAm, SAm
Canis gezi Kraglievich, 1928	E–M Pleist	SAm
Canis nehringi Ameghino, 1902	L Pleist	SAm
Canis etruscus Forsyth-Major, 1877	E Pleist	Eu
Canis falconeri Forsyth-Major 1877	E Pleist	Eu
Canis arnensis Del Campana, 1913	E Pleist	Eu
Canis antonii Zdansky, 1924	E Pleist	As
Canis chihliensis Zdansky, 1924	E Pleist	As
Canis palmidens (Teilhard and Piveteau, 1930)	L Plio	As
Canis variabilis Pei, 1934	M Pleist	As
Canis teilhardi Qiu, Deng, and Wang, 2004	L Plio	As
Canis longdanensis Qiu, Deng, and Wang, 2004	L Plio	As
Canis brevicephalus Qiu, Deng, and Wang, 2004	L Plio	As
Canis mosbachensis Soergel, 1925	M Pleist	Eu
Canis lupus Linnaeus, 1758, gray wolf	Pleist–R	As, Eu, NAm
Canis africanus Pohle, 1928	E Pleist	Af
Canis adustus Sundevall, 1847, side-striped jackal	R	Af
Canis aureus Linnaeus, 1758, golden jackal	R	Af, As
Canis mesomelas Schreber, 1775, black-backed jackal	E Pleist–R	Af

Canis simensis Rüppell, 1835, Ethiopian wolf	R	Af
Xenocyon dubius (Teilhard de Chardin, 1940)	M Pleist	As
Xenocyon texanus Troxell, 1915	L Pleist	NAm
Xenocyon lycaonoides Kretzoi, 1938	Pleist	As, Eu, Af, NAm
Cynotherium sardous Studiati, 1857	L Pleist	Eu
Cuon alpinus (Pallas, 1811), dhole	Pleist–R	As, Eu
Lycaon magnus Ewer and Singer, 1956	M Pleist	Af
Lycaon pictus (Temminck, 1820), African wild dog	L Pleist–R	Af

APPENDIX 2
PHYLOGENETIC TREE OF THE FAMILY CANIDAE

ONLY GENERIC-LEVEL RELATIONSHIPS ARE SHOWN, followed by the number of species in parentheses. The thick vertical bars indicate the approximate geologic range of each genus. The thin lines indicate proposed phyletic relationships. We follow Wang (1994) for the subfamily Hesperocyoninae; Wang, Tedford, and Taylor (1999) for the subfamily Borophaginae; and Tedford, Wang, and Taylor (1995, in press) and Berta (1988) for the subfamily Caninae.

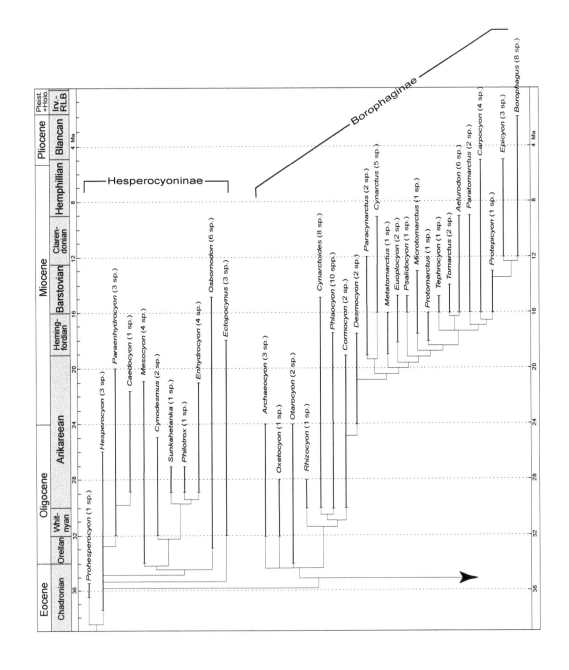

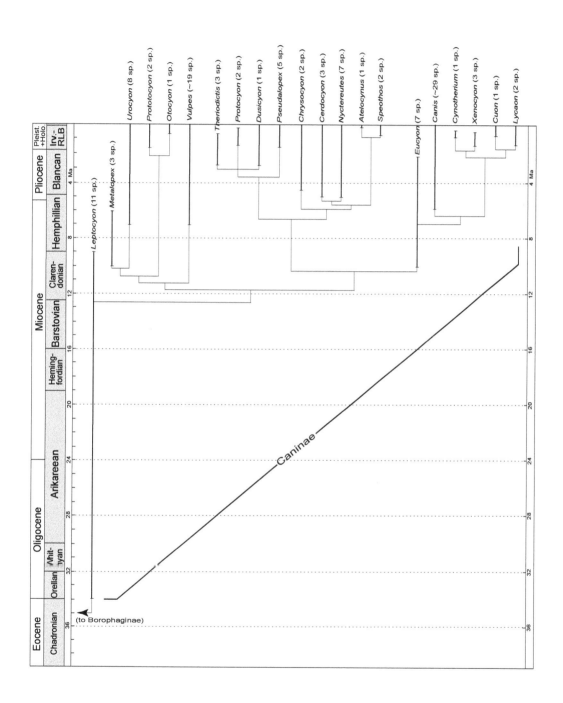

GLOSSARY

FAMILY AND SUBFAMILY NAMES are given in this glossary, but genus and species names are not included (for the latter, see the appendix and the index).

AMPHICYONIDS   Common name for the family Amphicyonidae, a group of extinct carnivorans best known for their possession of both doglike and bearlike characters; although commonly called bear dogs, amphicyonids are neither bears (ursids) nor dogs (canids).

ANAGENETIC CHANGE (ANAGENESIS)   The progression of evolution in which a lineage does not branch into different lineages, in contrast to *cladogenetic change*, in which the lineage branches into two or more lineages.

ARCTOIDS   Common name for the suborder Arctoidea, a diverse group of extinct and extant carnivorans that includes the dog family (Canidae), the bear family (Ursidae), aquatic carnivorans (Pinnipedia), the raccoon family (Procyonidae), the skunk family (Mephitidae), and the weasel family (Mustelidae).

ARTIODACTYLS   Common name for the order Artiodactyla, a diverse group of herbivorous mammals with an even number of hoofs, such as pigs, camels, deer, sheep, and cows.

AUDITORY BULLA   Bony enclosure in the mammalian ear region that protects the delicate ear bones (auditory ossicles).

AUDITORY OSSICLES   Three tiny bones (malleus, incus, and stapes) in the ears of mammals that transmit sound waves from the eardrum to the inner-ear nerves.

BIFURCATING PROCESS   *See* CLADOGENETIC CHANGE.

BORHYAENIDS   Common name for the family Borhyaenidae, a group of marsupial carnivores in the Miocene of South America.

BOROPHAGINES   Common name for the subfamily Borophaginae, a diverse group of extinct North American canids in the early Oligocene to the Pliocene.

BUNODONT   Form of dental pattern in which individual cusps are rounded and low crowned; bunodont teeth are often associated with an omnivorous diet.

CANIDS   Common name for the family Canidae, a group of carnivorans that includes the living wolves, foxes, coyotes, raccoon dogs, jackals, and hunting dogs, as well as their fossil relatives; canids have a worldwide distribution and their fossil ancestors date back to the late Eocene (more than 40 Ma).

CANIFORMS   Doglike carnivorans that include miacids, canids, amphicyonids, and arctoids; caniforms tend to have a primitively complete set of cheek teeth in contrast to the more reduced dental patterns in feliforms.

CANINES   Common name for the subfamily Caninae, a diverse group of extinct and extant canids from the early Oligocene to the present time; canines evolved in North America and migrated to Eurasia in the late Miocene and to South America in the Pliocene.

CANINE TOOTH   Mammalian tooth positioned between incisors and premolars; commonly long and sharp, canines are usually used as weapons by carnivorans.

CARNASSIALS   Pair of shearing teeth for processing meat formed by the upper fourth premolar and the lower first molar; all carnivorans possess a pair of carnassial teeth.

CARNIVORANS   Common name for the order Carnivora, a group of mammals that includes the living dogs, cats, raccoons, weasels, skunks, pandas, bears, and other lesser-known predators; most members of this order are predators and are defined by their possession of a pair of shearing teeth known as carnassials, made up of the upper fourth premolar and the lower first molar.

CARNIVORES   Animals that consume meat as their primary diet. Carnivores may or may not be members of the order Carnivora (or carnivorans) because some carnivorans are not meat eaters (such as the pandas) and some carnivores (such as hyaenodonts) are not carnivorans.

CENOZOIC   Geologic time period dominated by mammals that spans from 65 Ma to the present.

CERDOCYONINES   Common name for the subtribe Cerdocyonina, a group that includes the extant South American canids and their extinct relatives, some of which also lived in North America; cerdocyonines include various zorros, the maned wolf, and the bush dog.

CHARACTERS   Features in organisms that are used to study their genealogical relationships (for example, a large ear is a character for the fennec fox).

CIMOLESTIDS   Group of Mesozoic mammals that may be distantly related to carnivorans.

CLADE   Group of organisms with a common ancestor, often equal to a monophyletic group.

CLADISTICS   Scientific method to deduce phylogenetic relationships by searching for derived characters common to a subset of organisms.

CLADOGENETIC CHANGE (CLADOGENESIS)   The progression of evolution in which a lineage branches into two or more lineages, in contrast to *anagenetic change*, in which no branching occurs.

CLADOGRAM   Branched diagram of relationships based on the hypothesis of shared derived characters.

CONSERVATIVE   The tendency to retain ancestral conditions within a lineage.

CONVERGENCE   Evolutionary changes by independent lineages that result in similar appearances.

CREODONTS   Common name for the order Creodonta, a group of extinct predators mostly from the Paleogene that were eventually superseded by members of the Carnivora.

CRETACEOUS   Geologic time period that spans 145 to 65 Ma.

CURSORIALITY (CURSORIAL)   The ability to run fast and for a sustained period of time.

DIGITIGRADE   Standing posture in which the proximal parts of the hands and feet are lifted above the ground so that the animal stands on the digits.

ECTOTYMPANIC BONE   Portion of the auditory bulla that serves as the attachment site for the eardrum.

ENAMEL   The hard, white, shiny substance that coats the outer surface of mammalian teeth.

ENDEMIC   Organisms are referred as *endemic* if their natural distribution is restricted to a particular place or region.

ENTOTYMPANIC BONE   Portion of the auditory bulla that often forms the main chamber wall of the auditory bulla in carnivorans.

EOCENE   Geologic time period that spans 55 to 34 Ma.

FELIFORMS   Catlike carnivorans that include viverravids, nimravids (archaic saber-toothed animals), felids, hyaenids, viverrids, and herpestids; feliforms tend to have a reduced set of cheek teeth, in contrast to the more complete dental patterns in caniforms.

FRONTAL SINUS   Air space within the frontal bone, a character that has been repeatedly developed in various lineages of carnivorans.

HESPEROCYONINES   Common name for the subfamily Hesperocyoninae, a group of extinct North American canids from the late Eocene to the middle Miocene.

HOMOLOGY   Shared features among organisms that are due to their inheritance of these features from their common ancestor.

HYAENIDS   Common name for the family Hyaenidae, a group of carnivorans that includes the extant hyenas and aardwolves as well as their extinct relatives; hyaenids have lived in Eurasia, Africa, and North America from the Miocene to the present.

HYAENODONTS   Common name for the family Hyaenodontidae in the order Creodonta, a group of extinct dominant carnivores from the Paleogene to part of the Neogene.

HYPERCARNIVORY (HYPERCARNIVOROUS)   Dental adaptations toward efficient processing of meat, commonly involving sharp blades in the carnassial teeth for cutting flesh and tendons.

HYPOCARNIVORY (HYPOCARNIVOROUS)   Dental adaptations toward a more generalized feeding (de-emphasis of meat eating), commonly involving increased grinding areas in the posterior cheek teeth.

HYPOTHETICAL-DEDUCTIVE METHOD   The scientific philosophy that emphasizes the formulation of hypotheses and their logical corollaries as a primary means to advance science.

HYPSODONTY   Vertically elongated enamel surface (known as crown height) on the cheek teeth of herbivores; hypsodont teeth generally indicate a diet that contains tough fibers in vegetation (often grasses).

INCISOR   Frontmost row of teeth in mammals adjacent to the canines; incisors in carnivorans are often simple and chisel shaped.

INTERNAL CAROTID ARTERY   Small artery that goes through the auditory bulla and often forms the main supply of blood to the brain in carnivorans.

MESOCARNIVORY (MESOCARNIVOROUS)   Unspecialized teeth in carnivorans that are intermediate between the hypercarnivorous and hypocarnivorous extremes.

MESONYCHIDS   Common name for the family Mesonychidae, a group of extinct mammals from the Paleogene, some of which evolved predatory features similar to those of later carnivorans.

MESOZOIC   Geologic time period dominated by dinosaurs and other reptiles that spans 250 to 65 Ma.

MIACIDS   Common name for the archaic family Miacidae, a group of extinct caniform carnivorans from the Paleogene.

MIDDLE EAR   Region of the mammalian ear anatomy enclosed within the auditory bulla.

MIOCENE   Geologic time period that spans 24 to 5 Ma.

MOLARS   Permanent cheek teeth near the back of the jaws in mammals; molars are not replaced during life (that is, there is no milk tooth precursor, as there is for the premolar).

MONOPHYLY   Natural group of organisms that includes their common ancestors as well as all their descendants.

MUSTELIDS   Common name for the family Mustelidae, a group of arctoid carnivorans that includes the extant weasels, badgers, otters, martens, wolverines, and their extinct relatives.

NEOGENE   Geologic time period that includes the Miocene through the Pliocene (24 to 1.8 Ma).

OLIGOCENE   Geologic time period that spans 34 to 24 Ma.

OMNIVORES   Animals that consume a wide variety of food instead of concentrating on a single source of food, such as meat or vegetation, as their primary diets.

PALEOCENE   Geologic time period that spans 65 to 55 Ma.

PALEOGENE   Geologic time period that includes the Paleocene through the Oligocene (65 to 24 Ma).

PARSIMONY   The scientific methodology for choosing the simplest (the most parsimonious) hypothesis when faced with multiple alternative hypotheses; also known as "Ockham's razor," after the fourteenth-century English philosopher and Franciscan friar William of Ockham.

PHYLOGENETIC SYSTEMATICS   Studies of organismal classification and taxonomy based on the organisms' genealogical relationships.

PHYLOGENY   Genealogical relationships that attempt to reconstruct historical relationships among various lineages of organisms.

PLANTIGRADE   Standing posture in which the entire hand and foot touch the ground and weight is mostly spread on the palm of the hand and the heel of the foot.

PLATE TECTONICS   The study of large-scale configuration and movement of earth's lithosphere (continents and seas).

PLEISTOCENE   Geologic time period that spans 1.8 to 0.01 Ma, also popularly known as the Ice Age.

PLIOCENE   Geologic time period that spans 5 to 1.8 Ma.

PREMOLARS   Cheek teeth near the front of the jaws in mammals; premolars consist of deciduous (milk) premolars and permanent premolars that replace the deciduous teeth during life.

PROCYONIDS   Common name for the family Procyonidae, a group of arctoid carnivorans that includes the extant raccoons, kinkajous, ring tails, coatis, and olingos, as well as their extinct relatives.

RADIATION   The rapid expansion of diversity within a group of organisms, often in response to environmental changes or new resources.

RANCHO LA BREA   A series of tar pits in Hancock Park in the city of Los Angeles that has preserved one of the richest late Pleistocene mammal fauna in the world, including the highest concentration of dire wolves.

SELENODONT   Form of dental pattern often seen in the even-hoofed mammals (artiodactyls), in which crescent-shaped ridges are repeated; selenodont teeth are often associated with an herbivorous diet.

SEXUAL DIMORPHISM   The differential development of a morphological feature by males and females in the same species; markedly dimorphic features in some carnivorans include the size of the body and the length of the canine teeth.

STRATIGRAPHY   Geologic study of rocks according to their relative positions (for example, upper strata are younger than lower strata).

TALONIDS   Rear half of the lower cheek teeth (premolars and molars); in carnivorans, talonids often serve as a grinding part of the dentition.

TERRESTRIAL   Attributes related to living on solid ground (as opposed to, for example, aquatic, or on the water).

THYLACINIDS Group of Australian marsupial predators that includes the recently extinct Tasmanian wolf.

UNGULIGRADE Standing posture in many mammalian herbivores, such as horses, in which the entire hand and foot are lifted above ground, and the animals stand on the tip of their fingers and toes.

URSIDS Common name for the family Ursidae, a group of arctoid carnivorans that includes the extant bears and pandas as well as their extinct relatives.

VERTEBRATES Common name for the phylum or subphylum Vertebrata, which includes all animals with a backbone (vertebral column).

VIVERRAVIDS Common name for the archaic family Viverravidae, a group of extinct feliform carnivorans from the Paleogene that have a reduced set of cheek teeth.

ZOOGEOGRAPHY The study of animal distribution, dispersal, and migration.

FURTHER READING

THIS READING LIST OF PRIMARY LITERATURE for different topics is intended for serious readers to examine a particular subject further. It includes but also goes much beyond the literature cited in the text and provides a reasonably comprehensive list without delving too extensively into the details of historical treatments, such as living and fossil canid taxonomy.

ANATOMY, LOCOMOTION, AND FUNCTION

Andersson, K. 2004. Elbow-joint morphology as a guide to forearm function and foraging behaviour in mammalian carnivores. *Zoological Journal of the Linnean Society* 142:91–104.

Andersson, K., and L. Werdelin. 2003. The evolution of cursorial carnivores in the Tertiary: Implications of elbow-joint morphology. *Biology Letters* 270:163–165.

Antón, M., and A. Galobart 1999. Neck function and predatory behaviour in the scimitar tooth cat *Homotherium latidens* (Owen). *Journal of Vertebrate Paleontology* 19:771–784.

Baker, M. A., and L. W. Chapman. 1977. Rapid brain cooling in exercising dogs. *Science* 195:781–783.

Binder, W. J., and B. Van Valkenburgh. 2000. Development of bite strength and feeding behaviour in juvenile spotted hyenas (*Crocuta crocuta*). *Journal of Zoology* 252:273–283.

Carbone, C., G. M. Mace, S. C. Roberts, and D. W. Macdonald. 1999. Energetic constraints on the diet of terrestrial carnivores. *Nature* 402:286–288.

Carrano, M. T. 1997. Morphological indicators of foot posture in mammals: A statistical and biomechanical analysis. *Zoological Journal of the Linnean Society* 121:77–104.

Evans, H. E., and G. C. Christensen. 1979. *Miller's Anatomy of the Dog*. Philadelphia: Saunders.

Ewer, R. F., and R. Singer. 1956. Fossil Carnivora from Hopefield. *Annals of the South African Museum* 42:335–347.

Flower, W. H. 1869. On the value of the characters of the base of the cranium in the classification of the order Carnivora, and on the systematic position of *Bassaris* and other disputed forms. *Proceedings of the Zoological Society of London* 1869:4–37.

Gaspard, M. 1964. La région de l'angle mandibulaire chez les Canidae. *Mammalia* 28:249–329.

Hildebrand, M. 1952. An analysis of body proportions in the Canidae. *American Journal of Anatomy* 90:217–256.

Hildebrand, M. 1954. Comparative morphology of the body skeleton in recent Canidae. *University of California Publications in Zoology* 52:399–470.

Holliday, J. A., and S. J. Steppan. 2004. Evolution of hypercarnivory: The effect of specialization on morphological and taxonomic diversity. *Paleobiology* 30:108–128.

Hunt, R. M., Jr. 2003. Intercontinental migration of large mammalian carnivores: Earliest occurrence of the Old World beardog *Amphicyon* (Carnivora, Amphicyonidae) in North America. *Bulletin of the American Museum of Natural History* 279:77–115.

Lee, D. V., J. E. A. Bertram, and R. J. Todhunter. 1999. Acceleration and balance in trotting dogs. *Journal of Experimental Biology* 202:3565–3573.

Munthe, K. 1989. The skeleton of the Borophaginae (Carnivora, Canidae): Morphology and function. *University of California Publications in Geological Sciences* 133:1–115.

Newman, C., C. D. Buesching, and J. O. Wolff. 2005. The function of facial masks in "midguild" carnivores. *Oikos* 108:623–633.

Ortolani, A. 1999. Spots, stripes, tail tips, and dark eyes: Predicting the function of carnivore colour patterns using the comparative method. *Biological Journal of the Linnean Society* 67:433–476.

Rensberger, J. M. 1995. Determination of stresses in mammalian dental enamel and their relevance to the interpretation of feeding behaviors in extinct taxa. In J. J. Thomason, ed., *Functional Morphology in Vertebrate Paleontology*, 151–172. Cambridge: Cambridge University Press.

Rensberger, J. M. 1997. Mechanical adaptation in enamel. In W. V. Koenigswald and P. M. Sander, eds., *Tooth Enamel Microstructure*, 237–257. Rotterdam: Balkema.

Stefen, C. 1999. Enamel microstructure of recent and fossil Canidae (Carnivora: Mammalia). *Journal of Vertebrate Paleontology* 19:576–587.

Usherwood, J. R., and A. M. Wilson. 2005. Biomechanics: No force limit on greyhound sprint speed. *Nature* 438:753.

Van Valkenburgh, B., and W. J. Binder. 2000. Biomechanics and feeding behavior in carnivores: Comparative and ontogenetic studies. In P. Domenici and R. W. Blake, eds., *Biomechanics in Animal Bahaviour*, 223–235. Oxford: BIOS.

Van Valkenburgh, B., and K.-P. Koepfli. 1993. Cranial and dental adaptations to predation in canids. In N. Dunstone and M. L. Gorman, eds., *Mammals as Predators*, 15–37. Oxford: Oxford University Press.

Van Valkenburgh, B., T. Sacco, and X. Wang. 2003. Pack hunting in Miocene borophagine dogs: Evidence from craniodental morphology and body size. *Bulletin of the American Museum of Natural History* 279:147–162.

Van Valkenburgh, B., J. Theodor, A. Friscia, A. Pollack, and T. Rowe. 2004. Respiratory turbinates of canids and felids: A quantitative comparison. *Journal of Zoology* 264:281–293.

Wang, X. 1993. Transformation from plantigrady to digitigrady: Functional morphology of locomotion in *Hesperocyon* (Canidae: Carnivora). *American Museum Novitates* 3069:1–23.

Wang, X., and B. M. Rothschild. 1992. Multiple hereditary osteochondromata of Oligocene *Hesperocyon* (Carnivora: Canidae). *Journal of Vertebrate Paleontology* 12:387–394.

Werdelin, L. 1989. Constraint and adaptation in the bone-cracking canid *Osteoborus* (Mammalia: Canidae). *Paleobiology* 15:387–401.

White, T. E. 1954. Preliminary analysis of the fossil vertebrates of the Canyon Ferry Reservoir area. *Proceedings of the United States National Museum* 103:395–438.

Wöhrmann-Repenning, A. 1993. The anatomy of the vermeronasal complex of the fox (*Vulpes vulpes* [L.]) under phylogenetic and functional aspects. *Zoologische Jahrbucher: Abteilung für Anatomie und Ontogenie der Tiere* 123:353–361.

Wroe, S., C. McHenry, and J. Thomason. 2005. Bite club: Comparative bite force in big biting mammals and the prediction of predatory behaviour in fossil taxa. *Proceedings of the Royal Society of London*, Series B 272:619–625.

BODY WEIGHT ESTIMATES

Andersson, K. 2004. Predicting carnivoran body mass from a weight-bearing joint. *Journal of Zoology* 262:161–172.

Anyonge, W., and C. Roman. 2006. New body mass estimates for *Canis dirus*, the extinct Pleistocene dire wolf. *Journal of Vertebrate Paleontology* 26:209–212.

Kaufman, J. A., and R. J. Smith. 2002. Statistical issues in the prediction of body mass for Pleistocene canids. *Lethaia* 35:32–34.

Van Valkenburgh, B. 1990. Skeletal and dental predictors of body mass in carnivores. In J. Damuth and B. MacFadden, eds., *Body Size in Mammalian Paleobiology: Estimation and Biological Implications*, 181–205. Cambridge: Cambridge University Press.

Van Valkenburgh, B., and K.-P. Koepfli. 1993. Cranial and dental adaptations to predation in canids. In N. Dunstone and M. L. Gorman, eds., *Mammals as Predators*, 15–37. Oxford: Oxford University Press.

DOMESTICATION

Bardeleben, C., R. L. Moore, and R. K. Wayne. 2005. Isolation and molecular evolution of the Selenocysteine tRNA (Cf TRSP) and RNase P RNA (Cf RPPH1) genes in the dog family, Canidae. *Molecular Biology and Evolution* 22:347–359.

Cohn, J. 1997. How wild wolves became domestic dogs. *BioScience* 47:725–728.

Coppinger, R., and L. Coppinger. 2001. *Dogs: A Startling New Understanding of Canine Origin, Behavior, and Evolution*. New York: Scribner.

Davis, S. J. M., and F. R. Valla. 1978. Evidence for the domestication of the dog 12,000 years ago in the Natufian of Israel. *Nature* 276:608–610.

Dayan, T. 1994. Early domesticated dogs of the Near East. *Journal of Archaeological Science* 21:633–640.

Diamond, J. 1997. *Guns, Germs, and Steel: The Fates of Human Societies*. New York: Norton.

Hare, B., M. Brown, C. Williamson, and M. Tomasello. 2002. The domestication of social cognition in dogs. *Science* 298:1634–1636.

Hemmer, H. 1990. *Domestication: The Decline of Environmental Appreciation*. Cambridge: Cambridge University Press.

Leonard, J. A., R. K. Wayne, J. Wheeler, R. Valadez, S. Guillén, and C. Vilà. 2002. Ancient DNA evidence for Old World origin of New World dogs. *Science* 298:1613–1616.

Lindblad-Toh, K., C. M. Wade, T. S. Mikkelsen, E. K. Karlsson, D. B. Jaffe, M. Kamal, M. Clamp, J. L. Chang, E. J. Kulbokas, M. C. Zody, E. Mauceli, X. Xie, M. Breen, R. K. Wayne, E. A. Ostrander, C. P. Ponting, F. Galibert, D. R. Smith, P. J. deJong, E. Kirkness, P. Alvarez, T. Biagi, W. Brockman, J. Butler, C.-W. Chin, A. Cook, J. Cuff, M. J. Daly, D. DeCaprio, S. Gnerre, M.

Grabherr, M. Kellis, M. Kleber, C. Bardeleben, L. Goodstadt, A. Heger, C. Hitte, L. Kim, K.-P. Koepfli, H. G. Parker, J. P. Pollinger, S. M. J. Searle, N. B. Sutter, R. Thomas, C. Webber, and E. S. Lander. 2005. Genome sequence, comparative analysis, and haplotype structure of the domestic dog. *Nature* 438:803–819.

Morey, D. E. 1994. The early evolution of the domestic dog. *American Scientist* 82:336–347.

Nobis, G. 1979. Der älteste Haushund lebte vor 14000 Jahren. *Die Umschau* 19:610.

Olsen, S. J. 1985. *Origins of the Domestic Dog: The Fossil Record*. Tucson: University of Arizona Press.

Olsen, S. J., and J. W. Olsen. 1977. The Chinese wolf, ancestor of New World dogs. *Science* 197:533–535.

Parker, H. G., L. V. Kim, N. B. Sutter, S. Carlson, T. D. Lorentzen, T. B. Malek, G. S. Johnson, H. B. DeFrance, E. A. Ostrander, and L. Kruglyak. 2004. Genetic structure of the purebred domestic dog. *Science* 304:1160–1164.

Sablin, M. V., and G. A. Khlopachev. 2002. The earliest Ice Age dogs: Evidence from Eliseevichi. *Current Anthropology* 43:795–799.

Savolainen, P., T. Leitner, A. N. Wilton, E. Matisoo-Smith, and J. Lundeberg. 2004. A detailed picture of the origin of the Australian dingo, obtained from the study of mitochondrial DNA. *Proceedings of the National Academy of Sciences* 101:12387–12390.

Savolainen, P., Y. Zhang, J. Luo, J. Lundeberg, and T. Leitner. 2002. Genetic evidence for an East Asian origin of domestic dogs. *Science* 298:1610–1613.

Schleidt, W. M., and M. D. Shalter. 2004. Co-evolution of humans and canids, an alternative view of dog domestication: *Homo homini lupus? Evolution and Cognition* 9:57–72.

Trut, L. N. 1999. Early canid domestication: The farm-fox experiment. *American Scientist* 87:160–169.

Vilà, C., P. Savolainen, J. E. Maldonado, I. R. Amorim, J. E. Rice, R. L. Honeycutt, K. A. Crandall, J. Lundeberg, and R. K. Wayne. 1997. The domestic dog has an ancient and genetically diverse origin. *Science* 276:1687–1689.

EVOLUTION

Baskin, J. A. 1998. Evolutionary trends in the late Miocene hyena-like dog *Epicyon* (Carnivora, Canidae). In Y. Tomida, L. J. Flynn, and L. J. Jacobs, eds., *Advances in Vertebrate Paleontology and Geochronology*, 191–214. Tokyo: National Science Museum.

Cope, E. D. 1880. Extinct Batrachia. *American Naturalist* 14:609–610.

Dayan, M., D. Simberloff, E. Tchernov, and Y. Yom-Tov. 1992. Canine carnassials: Character displacement in the wolves, jackals, and foxes of Israel. *Biological Journal of the Linnean Society* 45:315–331.

Dayan, M., E. Tchernov, Y. Yom-Tov, and D. Simberloff. 1989. Ecological character displacement in Saharo-Arabian *Vulpes*: Outfoxing Bergmann's rule. *Oikos* 55:263–272.

Gittleman, J. L. 1986. Carnivore brain size, behavioral ecology, and phylogeny. *Journal of Mammalogy* 67:23–36.

Holliday, J. A., and S. J. Steppan. 2004. Evolution of hypercarnivory: The effect of specialization on morphological and taxonomic diversity. *Paleobiology* 30:108–128.

Kleiman, D. G., and J. F. Eisenberg. 1973. Comparisons of canid and felid social systems from an evolutionary perspective. *Animal Behavior* 21:637–659.

Muñoz-Durán, J. 2002. Correlates of speciation and extinction rates in the Carnivora. *Evolutionary Ecology Research* 4:963–991.

Radinsky, L. 1969. Outlines of canid and felid brain evolution. *Annals of the New York Academy of Sciences* 167:277–288.

Radinsky, L. 1973. Evolution of the canid brain. *Brain Behavior and Evolution* 7:169–202.

Turner, A., and M. Antón. 1996. *The Big Cats and Their Fossil Relatives*. New York: Columbia University Press.

Van Valkenburgh, B. 1991. Iterative evolution of hypercarnivory in canids (Mammalia: Carnivora): Evolutionary interactions among sympatric predators. *Paleobiology* 17:340–362.

Van Valkenburgh, B., and F. Hertel. 1993. Tough times at La Brea: Tooth breakage in large carnivores of the late Pleistocene. *Science* 261:456–459.

Werdelin, L., and M. E. Lewis. 2005. Plio-Pleistocene Carnivora of eastern Africa: Species richness and turnover patterns. *Zoological Journal of the Linnean Society* 144:121–144.

Wesley-Hunt, G. D., and J. J. Flynn. 2005. Phylogeny of the Carnivora: Basal relationships among the carnivoramorphans and assessment of the position of "Miacoidea" relative to Carnivora. *Journal of Systematic Palaeontology* 3:1–28.

FOSSIL SPECIES

Ballesio, R., and M. Philippe. 1995. Les Canidés Pléistocenes de la Balme a Collomb (commune d'Entremont-le-Vieux, Savoie). *Nouvelles Archives du Muséum d'Histoire Naturelle de Lyon* 33:45–65.

Barbour, E. H., and H. J. Cook. 1914. Two new fossil dogs of the genus *Cynarctus* from Nebraska. *Nebraska Geological Survey* 4:225–227.

Barbour, E. H., and H. J. Cook. 1917. Skull of *Aelurodon platyrhinus* sp. nov. *Nebraska Geological Survey* 7:173–180.

Barbour, E. H., and C. B. Schultz. 1935. A new Miocene dog, *Mesocyon geringensis* sp. nov. *Bulletin of the Nebraska State Museum* 1:407–418.

Baskin, J. A. 1980. The generic status of *Aelurodon* and *Epicyon* (Carnivora, Canidae). *Journal of Paleontology* 54:1349–1351.

Baskin, J. A. 2005. Carnivora from the late Miocene Love Bone Bed of Florida. *Bulletin of the Florida Museum of Natural History* 45:413–434.

Berry, C. T. 1938. A Miocene dog from Maryland. *Proceedings of the United States National Museum* 85:159–161.

Berta, A. 1981. Evolution of large canids in South America. *Anais do II Congreso Latino-Americano de Paleontologia, Porto Alegre* 2:835–845.

Berta, A. 1984. The Pleistocene bush dog *Speothos pacivorus* (Canidae) from the Lagoa Santa caves, Brazil. *Journal of Mammalogy* 65:549–559.

Berta, A. 1988. Quaternary evolution and biogeography of the large South American Canidae (Mammalia: Carnivora). *University of California Publications in Geological Sciences* 132:1–149.

Bever, G. S. 2005. Morphometric variation in the cranium, mandible, and dentition of *Canis latrans* and *Canis lepophagus* (Carnivora: Canidae) and its implications for the identification of isolated fossil specimens. *Southwestern Naturalist* 50:42–56.

Bjork, P. R. 1970. The Carnivora of the Hagerman Local Fauna (late Pliocene) of southwestern Idaho. *Transactions of the American Philosophical Society*, n.s., 60:1–54.

Bryant, H. N. 1991. Reidentification of the Chadronian supposed didelphid marsupial *Alloeodectes mcgrewi* as part of the deciduous dentition of the canid *Hesperocyon*. *Canadian Journal of Earth Sciences* 28:2062–2065.

Bryant, H. N. 1992. The Carnivora of the Lac Pelletier Lower Fauna (Eocene: Duchesnean), Cypress Hills Formation, Saskatchewan. *Journal of Paleontology* 66:847–855.

Bryant, H. N. 1993. Carnivora and Creodonta of the Calf Creek Local Fauna (late Eocene, Chadronian), Cypress Hills Formation, Saskatchewan. *Journal of Paleontology* 67:1032–1046.

Clark, J. 1939. *Miacis gracilis*, a new carnivore from the Unita Eocene (Utah). *Annals of the Carnegie Museum* 27:349–370.

Cook, H. J. 1909. Some new Carnivora from the lower Miocene beds of western Nebraska. *Nebraska Geological Survey* 3:262–272.

Cook, H. J. 1914. A new canid from the lower Pliocene of Nebraska. *Nebraska Geological Survey* 7:49–50.

Cook, H. J., and J. R. Macdonald. 1962. New Carnivora from the Miocene and Pliocene of western Nebraska. *Journal of Paleontology* 36:560–567.

Cope, E. D. 1873. Third notice of extinct Vertebrata from the Tertiary of the plains. *Paleontological Bulletin* 16:1–8.

Cope, E. D. 1877. Report upon the extinct Vertebrata obtained in New Mexico by parties of the expedition of 1874. In *Report upon United States Geological Surveys West of the One Hundredth Meridian, First Lieut. Geo. M. Wheeler, Corps of Engineers, U.S. Army, in Charge*, 4:1–370. Washington, D.C.: Government Printing Office.

Cope, E. D. 1883. On the extinct dogs of North America. *American Naturalist* 17:235–249.

Cope, E. D. 1884. The Vertebrata of the Tertiary formations of the West, book I. In *Report upon United States Geological Surveys West of the One Hundredth Meridian, Part II*, 3:1–1009. Washington, D.C.: Government Printing Office.

Cope, E. D. 1890. A new dog from the Loup Fork Miocene. *American Naturalist* 24:1067–1068.

Crusafont-Pairó, M. 1950. El primer representante del género *Canis* en el Pontiense eurasiatico (*Canis cipio* nova sp.). *Boletín de la Real Sociedad Española de Historia Natural (Geología)* 48:43–51.

Dayan, T. 1994. Carnivore diversity in the late Quaternary of Israel. *Quaternary Research* 41:343–349.

de Bonis, L., S. Peigné, A. Likius, H. T. Mackaye, P. Vignaud, and M. Brunet. 2007. The oldest African fox (*Vulpes riffautae* n. sp., Canidae, Carnivora) recovered in late Miocene deposits of the Djurab desert, Chad. *Naturwissenschaften* 94:575–580.

Emry, R. J., and R. E. Eshelman. 1998. The early Hemingfordian (early Miocene) Pollack Farm Local Fauna: First Tertiary land mammals described from Delaware. In R. N. Benson, ed., *Geology and Paleontology of the Lower Miocene Pollack Farm Fossil Site, Delaware*, 153–173. Delaware Geological Survey Special Publication no. 21. Newark: Delaware Geological Survey.

Evander, R. L. 1986. Carnivores of the Railway Quarries Local Fauna. *Transactions of the Nebraska Academy of Science* 14:25–34.

Eyerman, J. 1894. Preliminary notice of a new species of *Temnocyon* and a new genus from the John Day Miocene of Oregon. *American Geologist* 14:320–321.

Eyerman, J. 1896. The genus *Temnocyon* and a new species thereof and the new genus *Hypotemnodon*, from the John Day Miocene of Oregon. *American Geologist* 17:267–286.

Frailey, D. 1978. An early Miocene (Arikareean) fauna from north central Florida (the SB-1A Local Fauna). *Occasional Papers of the Museum of Natural History, University of Kansas* 75:1–20.

Frailey, D. 1979. The large mammals of the Buda Local Fauna (Arikareean: Alachua County, Florida). *Bulletin of the Florida State Museum, Biological Sciences* 24:123–173.

Galbreath, E. C. 1953. A contribution to the Tertiary geology and paleontology of northeastern Colorado. *University of Kansas Paleontological Contributions* 4:1–120.

Galbreath, E. C. 1956. Remarks on *Cynarctoides arcidens* from the Miocene of northeastern Colorado. *Transactions of the Kansas Academy of Science* 59:373–378.

Gawne, C. E. 1975. Rodents from the Zia Sand Miocene of New Mexico. *American Museum Novitates* 2586:1–25.

Gazin, C. L. 1932. A Miocene mammalian fauna from southeastern Oregon. *Carnegie Institution of Washington, Contributions to Paleontology* 418:37–86.

Gazin, C. L. 1942. The late Cenozoic vertebrate faunas from the San Pedro Valley, Ariz. *Proceedings of the United States National Museum* 92:475–518.

Geraads, D. 1997. Carnivores du Pliocène terminal de Ahl al Oughlam (Casablanca, Maroc). *Geobios* 30:127–164.

Green, M. 1948. A new species of dog from the lower Pliocene of California. *University of California Publications, Bulletin of the Department of Geological Sciences* 28:81–90.

Green, M. 1954. A cynarctine from the upper Oligocene of South Dakota. *Transactions of the Kansas Academy of Science* 57:218–220.

Gustafson, E. P. 1986. Carnivorous mammals of the late Eocene and early Oligocene of Trans-Pecos Texas. *Bulletin of the Texas Memorial Museum* 33:1–66.

Hall, E. R., and W. W. Dalquest. 1962. A new doglike carnivore, genus *Cynarctus*, from the Clarendonian, Pliocene, of Texas. *University of Kansas Publications, Museum of Natural History* 14:137–138.

Harrison, J. A. 1983. The Carnivora of the Edson Local Fauna (late Hemphillian), Kansas. *Smithsonian Contributions to Paleobiology* 54:1–42.

Hayes, F. G. 2000. The Brooksville 2 Local Fauna (Arikareean, latest Oligocene): Hernando County, Florida. *Bulletin of the Florida Museum of Natural History* 43:1–47.

Henshaw, P. C. 1942. A Tertiary mammalian fauna from the San Antonio Mountains near Tonopah, Nevada. *Carnegie Institution of Washington, Contributions to Paleontology* 530:77–168.

Hesse, C. J. 1936. A Pliocene vertebrate fauna from Optima, Oklahoma. *University of California Publications, Bulletin of the Department of Geological Sciences* 24:57–70.

Hibbard, C. W. 1950. Mammals of the Rexroad Formation from Fox Canyon, Meade County, Kansas. *Contributions from the Museum of Paleontology, University of Michigan* 8:113–192.

Hough, J. R., and R. Alf. 1956. A Chadron mammalian fauna from Nebraska. *Journal of Paleontology* 30:132–140.

Johnston, C. S. 1937. Tracks from the Pliocene of west Texas. *American Midland Naturalist* 28:147–152.

Johnston, C. S. 1939. A skull of *Osteoborus validus* from the early middle Pliocene of Texas. *Journal of Paleontology* 13:526–530.

Koufos, G. D. 1992. The Pleistocene carnivores of the Mygdonia basin (Macedonia, Greece). *Annales de Paléontologie* 78:205–257.

Koufos, G. D. 1997. The canids *Eucyon* and *Nyctereutes* from the Ruscinian of Macedonia, Greece. *Paleontologia i Evolució* 30–31:39–48.

Kurtén, B. 1974. A history of coyote-like dogs (Canidae, Mammalia). *Acta Zoologica Fennica* 140:1–38.

Kurtén, B. 1984. Geographic differentiation in the Rancholabrean dire wolf (*Canis dirus* Leidy) in North America. In H. H. Genoways and M. R. Dawson, eds., *Contributions in Quaternary Vertebrate Paleontology: A Volume in Memorial to John E. Guilday*, 218–227. Pittsburgh: Carnegie Museum of Natural History.

Leidy, J. 1858. Notice of remains of extinct Vertebrata, from the valley of the Niobrara River, collected during the exploring expedition of 1857, in Nebraska, under the command of Lieut. G. K. Warren, U.S. Top. Eng., by Dr. F. V. Hayden, geologist to the expedition. *Proceedings of the Academy of Natural Sciences of Philadelphia* 1858:20–29.

Leidy, J. 1869. The extinct mammalian fauna of Dakota and Nebraska, including an account of some allied forms from other localities, together with a synopsis of the mammalian remains of North America. *Journal of the Academy of Natural Sciences of Philadelphia* 7:1–472.

Loomis, F. B. 1931. A new Oligocene dog. *American Journal of Science* 22:100–102.

Loomis, F. B. 1932. The small carnivores of the Miocene. *American Journal of Science* 24:316–329.

Loomis, F. B. 1936. Three new Miocene dogs and their phylogeny. *Journal of Paleontology* 10:44–52.

Lyras, G. A., A. A. E. Van der Geer, M. D. Dermitzakis, and J. D. Vos. 2006. *Cynotherium sardous*, an insular canid (Mammalia: Carnivora) from the Pleistocene of Sardinia (Italy), and its origin. *Journal of Vertebrate Paleontology* 28:735–745.

Macdonald, J. R. 1948. The Pliocene carnivores of the Black Hawk Ranch Fauna. *University of California Publications, Bulletin of the Department of Geological Sciences* 28:53–80.

Macdonald, J. R. 1963. The Miocene faunas from the Wounded Knee area of western South Dakota. *Bulletin of the American Museum of Natural History* 125:141–238.

Macdonald, J. R. 1967a. A new species of late Oligocene dog, *Brachyrhynchocyon sesnoni*, from South Dakota. *Contributions in Science, Los Angeles County Museum* 126:1–5.

Macdonald, J. R. 1967b. A new species of late Oligocene dog, *Sunkahetanka sheffleri*, from South Dakota. *Contributions in Science, Los Angeles County Museum* 127:1–5.

Macdonald, J. R. 1970. Review of the Miocene Wounded Knee faunas of southwestern South Dakota. *Bulletin of the Los Angeles County Museum of Natural History* 8:1–82.

Martin, H. T. 1928. Two new carnivores from the Pliocene of Kansas. *Journal of Mammalogy* 9:233–236.

Martin, R. 1971. Les affinités de *Nyctereutes megamastoides* (Pomel) canidé du gisement Villafranchien de Saint-Vallier (Drôme, France). *Palaeovertebrata* 4:39–58.

Martin, R. 1973. Trois nouvelles espèces de Caninae (Canidae, Carnivora) des gisements plio-villafranchiens d'Europe. *Documents des Laboratoires de Géologie de la Faculté des Sciences de Lyon* 57:87–96.

Matthew, W. D. 1901. Fossil mammals of the Tertiary of northeastern Colorado. *Memoirs of the American Museum of Natural History* 1:355–448.

Matthew, W. D. 1902. New Canidae from the Miocene of Colorado. *Bulletin of the American Museum of Natural History* 16:281–290.

Matthew, W. D. 1918. Contributions to the Snake Creek fauna with notes upon the Pleistocene of western Nebraska, American Museum Expedition of 1916. *Bulletin of the American Museum of Natural History* 38:183–229.

Matthew, W. D. 1924. Third contribution to the Snake Creek fauna. *Bulletin of the American Museum of Natural History* 50:59–210.

McGrew, P. O. 1935. A new *Cynodesmus* from the lower Pliocene of Nebraska with notes on the phylogeny of dogs. *University of California Publications, Bulletin of the Department of Geological Sciences* 23:305–312.

McGrew, P. O. 1937. The genus *Cynarctus*. *Journal of Paleontology* 11:444–449.

McGrew, P. O. 1938. Dental morphology of the Procyonidae with a description of *Cynarctoides*, gen. nov. *Field Museum of Natural History, Geological Series* 6:323–339.

McGrew, P. O. 1941. A new procyonid from the Miocene of Nebraska. *Field Museum of Natural History, Geological Series* 8:33–36.

McGrew, P. O. 1944a. The *Aelurodon saevus* group. *Field Museum of Natural History, Geological Series* 8:79–84.

McGrew, P. O. 1944b. An *Osteoborus* from Honduras. *Field Museum of Natural History, Geological Series* 8:75–77.

McKenna, M. C., and S. K. Bell. 1997. *Classification of Mammals Above the Species Level*. New York: Columbia University Press.

Merriam, J. C. 1903. The Pliocene and Quaternary Canidae of the Great Valley of California. *University of California Publications, Bulletin of the Department of Geological Sciences* 3:277–290.

Merriam, J. C. 1906. Carnivora from the Tertiary formations of the John Day region. *University of California Publications, Bulletin of the Department of Geological Sciences* 5:1–64.

Merriam, J. C. 1911. Tertiary mammal beds of Virgin Valley and Thousand Creek in northwestern Nevada. *University of California Publications, Bulletin of the Department of Geological Sciences* 6:199–306.

Merriam, J. C. 1913. Notes on the canid genus *Tephrocyon*. *University of California Publications, Bulletin of the Department of Geological Sciences* 7:359–372.

Merriam, J. C. 1919. Tertiary mammalian faunas of the Mohave Desert. *University of California Publications, Bulletin of the Department of Biological Sciences* 11:437–585.

Merriam, J. C., and W. J. Sinclair. 1907. Tertiary faunas of the John Day region. *University of California Publications, Bulletin of the Department of Geological Sciences* 5:171–205.

Miller, W. E., and O. Carranza-Castañeda. 1998. Late Tertiary canids from central Mexico. *Journal of Paleontology* 72:546–556.

Morales, J., M. Pickford, and D. Soria. 2005. Carnivores from the late Miocene and basal Pliocene of the Tugen Hills, Kenya. *Revista de la Sociedad Geológica de España* 18:39–61.

Munthe, K. 1998. Canidae. In C. M. Janis, K. M. Scott, and L. L. Jacobs, eds., *Evolution of Tertiary Mammals of North America*, vol. 1, *Terrestrial Carnivores, Ungulates, and Ungulatelike Mammals*, 124–143. Cambridge: Cambridge University Press.

Nowak, R. M. 1979. North American Quaternary *Canis*. *Monograph of the Museum of Natural History, University of Kansas* 6:1–154.

Obara, I., and Y. Hasegawa. 2003. A skull of the Japanese wolf, *Canis hodophilax* Temminck, found in Ogura-yama limestone fissure, Ueno-mura, Gunma Prefecture. *Bulletin of the Gunma Museum of Natural History* 7:35–39.

Olsen, S. J. 1956a. The Caninae of the Thomas Farm Miocene. *Breviora* 26:1–12.

Olsen, S. J. 1956b. A new species of *Osteoborus* from the Bone Valley Formation of Florida. *Florida Geological Survey Special Publication* 2:1–5.

Olson, E. C., and P. O. McGrew. 1941. Mammalian fauna from the Pliocene of Honduras. *Bulletin of the Geological Society of America* 52:1219–1244.

Peterson, O. A. 1910. Description of new carnivores from the Miocene of western Nebraska. *Memoirs of the Carnegie Museum* 4:205–278.

Peterson, O. A. 1924. Discovery of fossil mammals in the Brown's Park Formation of Moffatt County, Colorado. *Annals of the Carnegie Museum* 15:299–304.

Prevosti, F. J., A. E. Zuritaa, and A. A. Carlini. 2005. Biostratigraphy, systematics, and paleoecology of *Protocyon* Giebel, 1855 (Carnivora, Canidae) in South America. *Journal of South American Earth Sciences* 20:5–12.

Qiu, Z., and R. H. Tedford. 1990. A Pliocene species of *Vulpes* from Yushe, Shanxi. *Vertebrata PalAsiatica* 28:245–258.

Richey, K. A. 1938. *Osteoborus diabloensis*, a new dog from the Black Hawk Ranch fauna, Mount Diablo, California. *University of California Publications, Bulletin of the Department of Geological Sciences* 24:303–307.

Richey, K. A. 1979. Variation and evolution in the premolar teeth of *Osteoborus* and *Borophagus* (Canidae). *Transactions of the Nebraska Academy of Science* 7:105–123.

Riggs, E. S. 1942. Preliminary description of two lower Miocene carnivores. *Field Museum of Natural History, Geological Series* 7:59–62.

Romer, A. S., and A. H. Sutton. 1927. A new arctoid carnivore from the lower Miocene. *American Journal of Science* 14:459–464.

Rook, L. 1992. "*Canis*" *monticinensis* sp. nov., a new Canidae (Carnivora, Mammalia) from the late Messinian of Italy. *Bolletino della Società Paleontologica Italiana* 31:151–156.

Rook, L. 1994. The Plio-Pleistocene Old World *Canis* (*Xenocyon*) ex gr. *falconeri*. *Bolletino della Società Paleontologica Italiana* 33:71–82.

Rook, L., and D. Torre. 1996a. The latest Villafranchian–early Galerian small dogs of the Mediterranean area. *Acta Zoologica Crucoviensia* 39:427–434.

Rook, L., and D. Torre. 1996b. The wolf-event in western Europe and the beginning of the late Villafranchian. *Neues Jahrbuch für Geologie und Paläontologie Abhandlungen* 1996:495–501.

Russell, L. S. 1934. Revision of the lower Oligocene vertebrate fauna of the Cypress Hills, Saskatchewan. *Transactions of the Royal Canadian Institute* 20:49–67.

Russell, R. D., and V. L. VanderHoof. 1931. A vertebrate fauna from a new Pliocene formation in northern California. *University of California Publications, Bulletin of the Department of Geological Sciences* 20:11–21.

Savage, D. E. 1941. Two new middle Pliocene carnivores from Oklahoma, with notes on the Optima fauna. *American Midland Naturalist* 25:692–710.

Schlaikjer, E. M. 1935. Contributions to the stratigraphy and paleontology of the Goshen Hole area, Wyoming. IV. New vertebrates and the stratigraphy of the Oligocene and early Miocene. *Bulletin of the Museum of Comparative Zoology* 76:97–189.

Scott, W. B. 1890a. The dogs of the American Miocene. *Princeton College Bulletin* 2:37–39.

Scott, W. B. 1890b. Preliminary account of the fossil mammals from the White River and Loup Fork formations, contained in the Museum of Comparative Zoology. Pt. II. The Carnivora and Artiodactyla. *Bulletin of the Museum of Comparative Zoology* 20:65–100.

Scott, W. B. 1893. The mammals of the Deep River beds. *American Naturalist* 27:659–662.

Scott, W. B. 1897. Preliminary notes on the White River Canidae. *Princeton University Bulletin* 9:1–3.

Scott, W. B. 1898. Notes on the Canidae of the White River Oligocene. *Transactions of the American Philosophical Society* 19:325–415.

Scott, W. B., and G. L. Jepsen. 1936. The mammalian fauna of the White River Oligocene. *Princeton University Bulletin* 28:1–980.

Sellards, E. H. 1916. Fossil vertebrates from Florida: A new Miocene fauna; new Pliocene species; the Pleistocene fauna. *Florida State Geological Survey Annual Report* 8:77–160.

Simpson, G. G. 1930. Tertiary land mammals of Florida. *Bulletin of the American Museum of Natural History* 59:149–211.

Simpson, G. G. 1932. Miocene land mammals from Florida. *Bulletin of the Florida State Geological Survey* 10:1–41.

Sinclair, W. J. 1915. Additions to the fauna of the lower Pliocene Snake Creek beds (results of the Princeton University 1914 expedition to Nebraska). *Proceedings of the American Philosophical Society* 54:73–95.

Soria, D., and E. Aguirre. 1976. El cánido de Layna: Revisión de los *Nyctereutes* fósiles. In M. T. Alberdi and E. Aguirre, eds., *Miscelanea neogena*, 83–116. Madrid: International Geological Correlation Program.

Sotnikova, M. V. 2001. Remains of Canidae from the lower Pleistocene site of Untermassfeld. In R.-D. Kahlke, ed., *Das Pleistozäne von Untermassfeld bei Meiningen (Thüringgen)*, pt. 2, 607–632. Mainz: Römisch-Germanischen Zentralmuseums.

Sotnikova, M. V. 2006. A new canid *Nurocyon chonokhariensis* gen. et sp. nov. (Canini, Canidae, Mammalia) from the Pliocene of Mongolia. *Courier Forschungsinstitut Senckenberg* 256:11–21.

Stevens, M. S. 1977. Further study of Castolon Local Fauna (early Miocene), Big Bend National Park, Texas. *Pearce-Sellards Series of the Texas Memorial Museum* 28:1–69.

Stevens, M. S. 1991. Osteology, systematics, and relationships of earliest Miocene *Mesocyon venator* (Carnivora: Canidae). *Journal of Vertebrate Paleontology* 11:45–66.

Stiner, M. C., F. C. Howell, B. Martinez-Navarro, E. Tchernov, and O. Bar-Yosef. 2001. Outside Africa: Middle Pleistocene *Lycaon* from Hayonim Cave, Israel. *Bolletino della Società Paleontologica Italiana* 40:293–302.

Stirton, R. A., and V. L. VanderHoof. 1933. *Osteoborus*, a new genus of dogs, and its relation to *Borophagus* Cope. *University of California Publications, Bulletin of the Department of Geological Sciences* 23:175–182.

Stock, C. 1928. Canid and proboscidian remains from the Ricardo deposits, Mohave Desert, California. *Carnegie Institution of Washington, Contributions to Paleontology* 393:39–49.

Stock, C. 1932. *Hyaenognathus* from the late Pliocene of the Coso Mountain, California. *Journal of Mammalogy* 13:263–266.

Stock, C. 1933. Carnivora from the Sespe of the Las Posas Hills, California. *Carnegie Institution of Washington, Contributions to Paleontology* 440:29–42.

Stock, C., and E. L. Furlong. 1926. New canid and rhinocerotid remains from the Ricardo Pliocene of the Mojave Desert, California. *University of California Publications, Bulletin of the Department of Biological Sciences* 16:43–60.

Tanner, L. G. 1973. Notes regarding skull characteristics of *Oxetocyon cuspidatus* Green (Mammalia, Canidae). *Transactions of the Nebraska Academy of Science* 2:66–69.

Tedford, R. H. 1978. History of dogs and cats: A view from the fossil record. In *Nutrition and Management of Dogs and Cats*, chap. M23. St. Louis: Ralston Purina.

Tedford, R. H. 1981. Mammalian biochronology of the late Cenozoic basins of New Mexico. *Bulletin of the Geological Society of America* 91:1008–1022.

Tedford, R. H., and D. Frailey. 1976. Review of some Carnivora (Mammalia) from the Thomas Farm local fauna (Hemingfordian: Gilchrist County, Florida). *American Museum Novitates* 2610:1–9.

Tedford, R. H., and Z. Qiu. 1991. Pliocene *Nyctereutes* (Carnivora: Canidae) from Yushe, Shanxi, with comments on Chinese fossil raccoon-dogs. *Vertebrata PalAsiatica* 29:176–189.

Tedford, R. H., and Z. Qiu. 1996. A new canid genus from the Pliocene of Yushe, Shanxi Province. *Vertebrata PalAsiatica* 34:27–40.

Tedford, R. H., and X. Wang. 2008. *Metalopex*, a new genus of fox (Carnivora: Canidae: Vulpini) from the late Miocene of western North America. *Contributions in Science, Natural History Museum of Los Angeles County* 41:273–278.

Tedford, R. H., X. Wang, and B. E. Taylor. In press. Phylogenetic systematics of the North American fossil Caninae (Carnivora: Canidae). *Bulletin of the American Museum of Natural History*.

Thorpe, M. R. 1922a. Oregon Tertiary Canidae, with descriptions of new forms. *American Journal of Science* 3:162–176.

Thorpe, M. R. 1922b. Some Tertiary Carnivora in the Marsh Collection, with descriptions of new forms. *American Journal of Science* 3:423–455.

VanderHoof, V. L. 1931. *Borophagus littoralis* from the marine Tertiary of California. *University of California Publications, Bulletin of the Department of Geological Sciences* 21:15–24.

VanderHoof, V. L. 1936. Notes on the type *Borophagus diversidens* Cope. *Journal of Mammalogy* 17:415–516.

VanderHoof, V. L., and J. T. Gregory. 1940. A review of the genus *Aelurodon*. *University of California Publications, Bulletin of the Department of Geological Sciences* 25:143–164.

Voorhies, M. R. 1965. The Carnivora of the Trail Creek Fauna. *Contributions to Geology, University of Wyoming* 4:21–25.

Wang, X. 1990. Pleistocene dire wolf remains from the Kansas River with notes on dire wolves in Kansas. *Occasional Papers of the Museum of Natural History, University of Kansas* 137:1–7.

Wang, X. 1994. Phylogenetic systematics of the Hesperocyoninae (Carnivora: Canidae). *Bulletin of the American Museum of Natural History* 221:1–207.

Wang, X. 2003. New material of *Osbornodon* from the early Hemingfordian of Nebraska and Florida. *Bulletin of the American Museum of Natural History* 279:163–176.

Wang, X., and R. H. Tedford. 1992. The status of genus *Nothocyon* Matthew, 1899 (Carnivora): An arctoid not a canid. *Journal of Vertebrate Paleontology* 12:223–229.

Wang, X., and R. H. Tedford. 1996. Canidae. In D. R. Prothero and R. J. Emry, eds., *The Terrestrial Eocene-Oligocene Transition in North America*, pt. 2, *Common Vertebrates of the White River Chronofauna*, 433–452. Cambridge: Cambridge University Press.

Wang, X., and R. H. Tedford. 2007. Evolutionary history of canids. In P. Jensen, ed., *The Behavioural Biology of Dogs*, 3–20. Oxford: CABI International.

Wang, X., and R. H. Tedford. 2008. Fossil dogs (Carnivora, Canidae) from the Sespe and Vaqueros formations in southern California, with comments on relationships of *Phlaocyon taylori*. *Contributions in Science, Natural History Museum of Los Angeles County* 41:255–272.

Wang, X., R. H. Tedford, and B. E. Taylor. 1999. Phylogenetic systematics of the Borophaginae (Carnivora: Canidae). *Bulletin of the American Museum of Natural History* 243:1–391.

Wang, X., R. H. Tedford, B. Van Valkenburgh, and R. K. Wayne. 2004a. Ancestry: Evolutionary history, molecular systematics, and evolutionary ecology of Canidae. In D. W. Macdonald and

C. Sillero-Zubiri, eds., *The Biology and Conservation of Wild Canids*, 39–54. Oxford: Oxford University Press.

Wang, X., R. H. Tedford, B. Van Valkenburgh, and R. K. Wayne. 2004b. Phylogeny, classification, and evolutionary ecology of the Canidae. In C. Sillero-Zubiri, M. Hoffmann, and D. W. Macdonald, eds., *Canids: Foxes, Wolves, Jackals, and Dogs. Status Survey and Conservation Action Plan*, 8–20. Gland, Switzerland: International Union for the Conservation of Nature and Natural Resources, Species Survival Commission, Canid Specialist Group, World Conservation Union.

Wang, X., B. C. Wideman, R. Nichols, and D. L. Hanneman. 2004. A new species of *Aelurodon* (Carnivora, Canidae) from the Barstovian of Montana. *Journal of Vertebrate Paleontology* 24:445–452.

Webb, S. D. 1969a. The Burge and Minnechaduza Clarendonian mammalian fauna of north-central Nebraska. *University of California Publications in Geological Sciences* 78:1–191.

Webb, S. D. 1969b. The Pliocene Canidae of Florida. *Bulletin of the Florida State Museum, Biological Sciences* 14:273–308.

Webb, S. D., B. J. MacFadden, and J. A. Baskin. 1981. Geology and paleontology of the Love Bone bed from the late Miocene of Florida. *American Journal of Science* 281:513–544.

Webb, S. D., and S. C. Perrigo. 1984. Late Cenozoic vertebrates from Honduras and El Salvador. *Journal of Vertebrate Paleontology* 4:237–254.

Werdelin, L., and M. E. Lewis. 2005. Plio-Pleistocene Carnivora of eastern Africa: Species richness and turnover patterns. *Zoological Journal of the Linnean Society* 144:121–144.

White, T. E. 1941. Additions to the fauna of the Florida Pliocene. *Proceedings of the New England Zoological Club* 18:67–70.

White, T. E. 1942. The lower Miocene mammal fauna of Florida. *Bulletin of the Museum of Comparative Zoology* 92:1–49.

White, T. E. 1947. Addition to the Miocene fauna of north Florida. *Bulletin of the Museum of Comparative Zoology* 99:497–515.

Wilson, J. A. 1939. A new species of dog from the Miocene of Colorado. *Contributions from the Museum of Paleontology, University of Michigan* 5:315–318.

Wilson, J. A. 1960. Miocene carnivores, Texas coastal plain. *Journal of Paleontology* 34:983–1000.

Wortman, J. L., and W. D. Matthew. 1899. The ancestry of certain members of the Canidae, the Viverridae, and Procyonidae. *Bulletin of the American Museum of Natural History* 12:109–139.

MODERN SPECIES

Ansorge, H. 1994. Intrapopula skull variability in the red fox, *Vulpes vulpes* (Mammalia: Carnivora: Canidae). *Zoologische Abhandlungen* 48:103–123.

Audet, A. M., C. B. Robbins, and S. Larivière. 2002. *Alopex lagopus*. *Mammalian Species* 713:1–10.

Beisiegel, B. D. M., and G. L. Zuercher. 2005. *Speothos venaticus*. *Mammalian Species* 783:1–6.

Bekoff, M. 1977. *Canis latrans*. *Mammalian Species* 79:1–9.

Berta, A. 1982. *Cerdocyon thous*. *Mammalian Species* 186:1–4.

Berta, A. 1986. *Aletocynus microtis*. *Mammalian Species* 256:1–3.

Cabrera, A. 1931. On some South America canine genera. *Journal of Mammalogy* 12:54–67.

Churcher, C. S. 1960. Cranial variation in the North American red fox. *Journal of Mammalogy* 41:349–360.

Clark, H. O. 2005. *Otocyon megalotis. Mammalian Species* 766:1–5.

Cohen, J. A. 1978. *Cuon alpinus. Mammalian Species* 100:1–3.

Dietz, J. M. 1985. *Chrysocyon brachyurus. Mammalian Species* 234:1–4.

Egoscue, H. J. 1979. *Vulpes velox. Mammalian Species* 122:1–5.

Frafjord, K., and I. Stevy. 1998. The red fox in Norway: Morphological adaptation or random variation in size? *Zeitschrift für Säugetierkunde* 63:16–25.

Fritzell, E. K., and K. J. Haroldson. 1982. *Urocyon cinereoargenteus. Mammalian Species* 198:1–8.

Fuller, T. K., and P. W. Kat. 1993. Hunting success of African wild dogs in southwestern Kenya. *Journal of Mammalogy* 74:464–467.

Fuller, T. K., T. H. Nicholls, and P. W. Kat. 1995. Prey and estimated food consumption of African wild dogs in Kenya. *South African Journal of Wildlife Research* 25:3.

Geffen, E., M. E. Gompper, J. L. Gittleman, H.-K. Luh, D. W. Macdonald, and R. K. Wayne. 1996. Size, life-history traits, and social organization in the Canidae: A reevaluation. *American Naturalist* 147:140–160.

Geffen, E., R. Hefner, D. W. Macdonald, and M. Ucko. 1992. Diet and foraging behavior of Blanford's foxes, *Vulpes cana*, in Israel. *Journal of Mammalogy* 73:395–402.

Gittleman, J. L. 1986. Carnivore brain size, behavioral ecology, and phylogeny. *Journal of Mammalogy* 67:23–36.

Gittleman, J. L., and S. L. Pimm. 1991. Crying wolf in North America. *Nature* 351:524–525.

Hall, E. R. 1981. *The Mammals of North America.* New York: Wiley.

Kolenosky, G. B., and R. O. Standfield. 1974. Morphological and ecological variation among gray wolves (*Canis lupus*) of Ontario, Canada. In M. W. Fox, ed., *The Wild Canids: Their Systematics, Behavioral Ecology, and Evolution*, 62–72. New York: Van Nostrand Reinhold.

Koler-Matznick, J., I. L. Brisbin Jr., M. Feinstein, and S. Bulmer. 2003. An updated description of the New Guinea singing dog (*Canis hallstromi*, Throughton 1957). *Journal of Zoology* 261:109–118.

Langguth, A. 1974. Ecology and evolution in the South American canids. In M. W. Fox, ed., *The Wild Canids: Their Systematics, Behavioral Ecology, and Evolution*, 192–206. New York: Van Nostrand Reinhold.

Larivière, S. 2002. *Vulpes zerda. Mammalian Species* 714:1–5.

Larivière, S., and M. Pasitschniak-Arts. 1996. *Vulpes vulpes. Mammalian Species* 537:1–11.

Larivière, S., and P. J. Seddon. 2001. *Vulpes rueppelli. Mammalian Species* 678:1–5.

Leonard, J. A., C. Vilà, and R. K. Wayne. 2005. Legacy lost: Genetic variability and population size of extirpated US grey wolves (*Canis lupus*). *Molecular Ecology* 14:9–17.

Macdonald, D. W., S. Creel, and M. G. L. Mills. 2004. Canid society. In D. W. Macdonald and C. Sillero-Zubiri, eds., *The Biology and Conservation of Wild Canids*, 85–106. Oxford: Oxford University Press.

Macdonald, D. W., and C. Sillero-Zubiri, eds. 2004. *The Biology and Conservation of Wild Canids.* Oxford: Oxford University Press.

MacIntosh, N. W. G. 1974. The origin of the dingo: An enigma. In M. W. Fox, ed., *The Wild Canids: Their Systematics, Behavioral Ecology, and Evolution*, 87–88. New York: Van Nostrand Reinhold.

Martensz, P. N. 1971. Observations on the food of the fox, *Vulpes vulpes* (L.), in an arid environment. *Wildlife Research* 16:73–75.

McGrew, J. C. 1979. *Vulpes macrotis. Mammalian Species* 123:1–6.

Mech, L. D. 1974. *Canis lupus. Mammalian Species* 37:1–6.

Mendelssohn, H., Y. Tom-Tov, G. Ilany, and D. Meninger. 1987. On the occurrence of Blanford's fox, *Vulpes cana* Blanford, 1877, in Israel and Sinai. *Mammalia* 51:459–462.

Mivart, St. G. 1890. *A Monograph of the Canidae: Dogs, Jackals, Wolves, and Foxes*. London: Porter and Dulau.

Moore, C. M., and P. W. Collins. 1995. *Urocyon littoralis*. *Mammalian Species* 489:1–7.

Novaro, A. J. 1997. *Pseudalopex culpaeus*. *Mammalian Species* 558:1–8.

Nowak, R. M. 1992. The red wolf is not a hybrid. *Conservation Biology* 6:593–595.

Nowak, R. M. 2002. The original status of wolves in eastern North America. *Southeastern Naturalist* 1:95–130.

O'Brien, S. J., and E. Mayr. 1991. Bureaucratic mischief: Recognizing endangered species and subspecies. *Science* 251:1187–1188.

Osgood, W. H. 1934. The genera and subgenera of South American canids. *Journal of Mammalogy* 15:45–50.

Paradiso, J. L., and R. M. Nowak. 1971. A report on the taxonomic status and distribution of the red wolf. *Fish and Wildlife Service Special Scientific Report* 145:1–36.

Paradiso, J. L., and R. M. Nowak. 1972. *Canis rufus*. *Mammalian Species* 22:1–4.

Peres, C. A. 1991. Observations on hunting by small-eared (*Atelocynus microtis*) and bush dogs (*Speothos venaticus*) in central–western Amazonia. *Mammalia* 55:635–639.

Rook, L., and M. L. A. Puccetti. 1996. Remarks on the skull morphology of the endangered Ethiopian jackal, *Canis simensis* Rüppel 1838. *Scienze Fisiche e Naturali* 7:277–302.

Rosenzweig, M. L. 1966. Community structure in sympatric Carnivora. *Journal of Mammalogy* 47:602–612.

Servin, J. I. 1991. Algunos aspectos de la conducta social del lobo mexicano (*Canis lupus baileyi*) en cautiverio. *Acta Zoologica Mexicana*, n.s., 45:1–43.

Servin, J. I., and C. Huxley. 1991. La dieta del coyote en un bosque de encino-pino de la Sierra Madre Occidental de Durango, Mexico. *Acta Zoologica Mexicana*, n.s., 44:1–26.

Servin, J. I., J. R. Rau, and M. Delibes. 1987. Use of radio tracking to improve the estimation by track counts of the relative abundance of red fox. *Acta Theriologica* 32:489–492.

Servin, J. I., J. R. Rau, and M. Delibes. 1991. Activity pattern of the red fox *Vulpes vulpes* in Donana, SW Spain. *Acta Theriologica* 36:369–373.

Sillero-Zubiri, C., and D. Gottelli. 1994. *Canis simensis*. *Mammalian Species* 485:1–6.

Stains, H. J. 1975. Calcanea of members of the Canidae. *Bulletin of the Southern California Academy of Sciences* 74:143–155.

Strahl, S. D., J. L. Silva, and I. R. Goldstein. 1992. The bush dog (*Speothos venaticus*) in Venezuela. *Mammalia* 56:9–13.

Walton, L. R., and D. O. Joly. 2003. *Canis mesomelas*. *Mammalian Species* 715:1–9.

Ward, O. G., and D. H. Wurster-Hill. 1990. *Nyctereutes procyonoides*. *Mammalian Species* 358:1–5.

Wayne, R. K. 1986. Cranial morphology of domestic and wild canids: The influence of development on morphological change. *Evolution* 40:243–261.

Wayne, R. K., S. B. George, D. Gilbert, P. W. Collins, S. D. Kovach, D. J. Girman, and N. Lehman. 1991. A morphologic and genetic study of the island fox, *Urocyon littoralis*. *Evolution* 45:1849–1868.

Wilson, P. J., S. Grewal, I. D. Lawford, J. N. M. Heal, A. G. Granacki, D. Pennock, J. B. Theberge, M. T. Theberge, D. R. Voigt, W. Waddell, R. E. Chambers, P. C. Paquet, G. Goulet, D. Cluff, and B. N. White. 2000. DNA profiles of the eastern Canadian wolf and the red wolf provide evidence for a common evolutionary history independent of the gray wolf. *Canadian Journal of Zoology* 78:2156–2166.

Wozencraft, W.C. 1993. Order Carnivora. In D.E. Wilson and D.M. Reeder, eds., *Mammal Species of the World: A Taxonomic and Geographic Reference*, 2d ed., 279–348. Washington, D.C.: Smithsonian Institution Press.

Yahnke, C.J., W.E. Johnson, E. Geffen, D. Smith, F. Hertel, M.S. Roy, C.F. Bonacic, T.K. Fuller, B. Van Valkenburgh, and R.K. Wayne. 1996. Darwin's fox: A distinct endangered species in a vanishing habitat. *Conservation Biology* 10:366–375.

Yeager, L.E. 1938. Tree-climbing by a gray fox. *Journal of Mammalogy* 19:376.

Zimmerman, R.S. 1938. A coyote's speed and endurance. *Journal of Mammalogy* 19:400.

MOLECULAR BIOLOGY

Bardeleben, C., R.L. Moore, and R.K. Wayne. 2005. Isolation and molecular evolution of the Selenocysteine tRNA (Cf TRSP) and RNase P RNA (Cf RPPH1) genes in the dog family, Canidae. *Molecular Biology and Evolution* 22:347–359.

Gottelli, D., C. Sillero-Zubiri, G.D. Applebaum, M.S. Roy, D.J. Girman, J. Garcia-Moreno, E.A. Ostrander, and R.K. Wayne. 1994. Molecular genetics of the most endangered canid: The Ethiopian wolf, *Canis simensis*. *Molecular Ecology* 3:277–290.

Lan, H., and L. Shi. 1996. The mitochondrial DNA evolution of four species of Canidae. *Acta Zoologica Sinica* 42:87–95.

Leonard, J. A., C. Vilà, K. Fox-Dobbs, P. L. Koch, R. K. Wayne, and B. Van Valkenburgh. 2007. Megafaunal extinctions and the disappearance of a specialized wolf Ecomorph. *Current Biology* 17:1146–1150.

Lindblad-Toh, K., C.M. Wade, T.S. Mikkelsen, E.K. Karlsson, D.B. Jaffe, M. Kamal, M. Clamp, J.L. Chang, E.J. Kulbokas, M.C. Zody, E. Mauceli, X. Xie, M. Breen, R.K. Wayne, E.A. Ostrander, C.P. Ponting, F. Galibert, D.R. Smith, P.J. deJong, E. Kirkness, P. Alvarez, T. Biagi, W. Brockman, J. Butler, C.-W. Chin, A. Cook, J. Cuff, M.J. Daly, D. DeCaprio, S. Gnerre, M. Grabherr, M. Kellis, M. Kleber, C. Bardeleben, L. Goodstadt, A. Heger, C. Hitte, L. Kim, K.-P. Koepfli, H.G. Parker, J.P. Pollinger, S.M.J. Searle, N.B. Sutter, R. Thomas, C. Webber, and E.S. Lander. 2005. Genome sequence, comparative analysis, and haplotype structure of the domestic dog. *Nature* 438:803–819.

Roy, M.S., E. Geffen, D. Smith, E.A. Ostrander, and R.K. Wayne. 1994. Patterns of differentiation and hybridization in North American wolflike canids, revealed by analysis of microsatellite loci. *Molecular Biology and Evolution* 11:553–570.

Wayne, R.K. 1993. Molecular evolution of the dog family. *Trends in Genetics* 9:218–224.

Wayne, R.K., E. Geffen, and C. Vilà. 2004. Population genetics: Population and conservation genetics of canids. In D.W. Macdonald and C. Sillero-Zubiri, eds., *The Biology and Conservation of Wild Canids*, 55–84. Oxford: Oxford University Press.

Wayne, R.K., and S.M. Jenks. 1991. Mitochondrial DNA analysis implying extensive hybridization of the endangered red wolf *Canis rufus. Nature* 351:565–568.

Waync, R.K., W.G. Nash, and S.J. O'Brien. 1987a. Chromosomal evolution of the Canidae. I. Species with high diploid numbers. *Cytogenetics and Cell Genetics* 44:123–133.

Wayne, R.K., W.G. Nash, and S.J. O'Brien. 1987b. Chromosomal evolution of the Canidae. II. Divergence from the primitive carnivore karyotype. *Cytogenetics and Cell Genetics* 44:134–141.

Wayne, R. K., B. Van Valkenburgh, P. W. Kat, T. K. Fuller, W. E. Johnson, and S. J. O'Brien. 1989. Genetic and morphological divergence among sympatric canids. *Journal of Heredity* 80:447–454.

PALEOCLIMATE AND PALEOECOLOGY

Berggren, W. A., and D. R. Prothero. 1992. Eocene–Oligocene climatic and biotic evolution. In D. R. Prothero and W. A. Berggren, eds., *Eocene–Oligocene Climatic and Biotic Evolution*, 1–28. Princeton, N.J.: Princeton University Press.

Cavelier, C., J.-J. Chateauneuf, C. Pomerol, D. Rabussier, M. Renard, and C. V. Grazzini. 1981. The geological events at the Eocene/Oligocene boundary. *Palaeogeography, Palaeoclimatology, Palaeoecology* 36:223–248.

Cerling, T. E. 1992. Development of grasslands and savannas in East Africa during the Neogene. *Palaeogeography, Palaeoclimatology, Palaeoecology* 97:241–247.

Cerling, T. E., J. Quade, Y. Wang, and J. R. Bowman. 1989. Carbon isotopes in soils and paleosols as ecology and paleoecology indicators. *Nature* 341:138–139.

Crusafont-Pairó, M., and J. Truyols-Santonja. 1956. A biometric study of the evolution of fissiped carnivores. *Evolution* 10:314–332.

DeConto, R. M., and D. Pollard. 2003. Rapid Cenozoic glaciation of Antarctica induced by declining atmospheric CO_2. *Nature* 421:245–249.

Fortelius, M., J. Eronen, J. Jernvall, L. Liu, D. Pushkina, J. Rinne, A. Tesakov, I. A. Vislobokova, Z. Zhang, and L. Zhou. 2002. Fossil mammals resolve regional patterns of Eurasian climate change over 20 million years. *Evolutionary Ecology Research* 4:1005–1016.

Fortelius, M., and N. Solounias. 2000. Functional characterization of ungulate molars using the abrasion-attrition wear gradient: A new method for reconstructing paleodiets. *American Museum Novitates* 3301:1–36.

Hu, Y., J. Meng, Y. Wang, and C. Li. 2005. Large Mesozoic mammals fed on young dinosaurs. *Nature* 433:149–152.

Hunt, R. M., Jr. 2004. Global climate and the evolution of large mammalian carnivores during the later Cenozoic in North America. *Bulletin of the American Museum of Natural History* 285:139–156. [Special issue: G. C. Gould and S. K. Bell, eds., *Tributes to Malcolm C. McKenna: His Students, His Legacy*]

Janis, C. M., J. Damuth, and J. M. Theodor. 2002. The origins and evolution of the North American grassland biome: The story from the hoofed mammals. *Palaeogeography, Palaeoclimatology, Palaeoecology* 177:183–198.

Kennett, J. P. 1977. Cenozoic evolution of Antarctic glaciation, the circum-Antarctic oceans and their impact on global paleoceanography. *Journal of Geophysical Research* 82:3843–3860.

Leopold, E. B., G. Liu, and S. Clay-Poole. 1992. Low-biomass vegetation in the Oligocene? In D. R. Prothero and W. A. Berggren, eds., *Eocene–Oligocene Climatic and Biotic Evolution*, 399–420. Princeton, N.J.: Princeton University Press.

MacFadden, B. J. 2000. Cenozoic mammalian herbivores from the Americas: Reconstructing ancient diets and terrestrial communities. *Annual Review of Ecology and Systematics* 31:33–59.

MacFadden, B. J., and T. E. Cerling. 1994. Fossil horses, carbon isotopes, and global change. *Trends in Ecology and Evolution* 9:481–486.

MacFadden, B. J., and T. E. Cerling. 1996. Mammalian herbivore communities, ancient feeding ecology, and carbon isotopes: A 10-million-year sequence from the Neogene of Florida. *Journal of Vertebrate Paleontology* 16:103–115.

Palmqvist, P., A. Arribas, and B. Martínez-Navarro. 1999. Ecomorphological study of large canids from the lower Pleistocene of southeastern Spain. *Lethaia* 32:75–88.

Palmqvist, P., B. Martínez-Navarro, and A. Arribas. 1996. Prey selection by terrestrial carnivores in a lower Pleistocene paleocommunity. *Paleobiology* 22:514–534.

Pearson, P. M., P. W. Ditchfield, J. Singano, K. G. Harcourt-Brown, C. J. Nicholas, R. K. Olsson, N. J. Shackleton, and M. A. Hall. 2001. Warm tropical sea surface temperatures in the late Cretaceous and Eocene epochs. *Nature* 413:481–487.

Prothero, D. R., and T. H. Heaton. 1996. Faunal stability during the early Oligocene climatic crash. *Palaeogeography, Palaeoclimatology, Palaeoecology* 127:257–283.

Quade, J., and T. E. Cerling. 1995. Expansion of C4 grasses in the late Miocene of northern Pakistan: Evidence from stable isotopes in paleosols. *Palaeogeography, Palaeoclimatology, Palaeoecology* 115:91–116.

Ramstein, G., F. Fluteau, J. Bese, and S. Joussaume. 1997. Effect of orogeny, plate motion, and land-sea distribution on Eurasian climate change over the past 30 million years. *Nature* 386:788–795.

Ridgway, K. D., A. R. Sweet, and A. R. Cameron. 1995. Climatically induced floristic changes across the Eocene–Oligocene transition in the northern high latitudes, Yukon Territory, Canada. *Bulletin of the Geological Society of America* 107:676–696.

Solounias, N., and G. Semprebon. 2002. Advances in the reconstruction of ungulate ecomorphology with application to early fossil equids. *American Museum Novitates* 3366:1–49.

Strömberg, C. A. E. 2002. The origin and spread of grass-dominated ecosystems in the late Tertiary of North America: Preliminary results concerning the evolution of hypsodonty. *Palaeogeography, Palaeoclimatology, Palaeoecology* 177:59–75.

Van Valkenburgh, B. 1994. Extinction and replacement among predatory mammals in the North American late Eocene–Oligocene: Tracking a guild over twelve million years. *Historical Biology* 8:1–22.

Van Valkenburgh, B. 1999. Major patterns in the history of carnivorous mammals. *Annual Review of Earth and Planetary Science* 27:463–493.

Van Valkenburgh, B. 2001. The dog-eat-dog world of carnivores: A review of past and present carnivore community dynamics. In C. Stanford and H. T. Bunn, eds., *Meat-Eating and Human Evolution*, 101–121. Oxford: Oxford University Press.

Van Valkenburgh, B., and F. Hertel. 1993. Tough times at La Brea: Tooth breakage in large carnivores of the late Pleistocene. *Science* 261:456–459.

Van Valkenburgh, B., and T. Sacco. 2002. Sexual dimorphism, social behavior, and intrasexual competition in large Pleistocene carnivorans. *Journal of Vertebrate Paleontology* 22:164–169.

Van Valkenburgh, B., T. Sacco, and X. Wang. 2003. Pack hunting in Miocene borophagine dogs: Evidence from craniodental morphology and body size. *Bulletin of the American Museum of Natural History* 279:147–162.

Van Valkenburgh, B., X. Wang, and J. Damuth. 2004. Cope's rule, hypercarnivory, and extinction in North American canids. *Science* 306:101–104.

Wang, Y., and T. E. Cerling. 1994. A model of fossil tooth and bone diagenesis: Implications for paleodiet reconstruction from stable isotopes. *Palaeogeography, Palaeoclimatology, Palaeoecology* 107:281–289.

Wang, Y., T. E. Cerling, and B. J. MacFadden. 1994. Fossil horses and carbon isotopes: New evidence for Cenozoic dietary, habitat, and ecosystem changes in North America. *Palaeogeography, Palaeoclimatology, Palaeoecology* 107:269–279.

Webb, S. D. 1977. A history of savanna vertebrates in the New World. Part I: North America. *Annual Review of Ecology and Systematics* 8:355–380.

Wing, S. L., G. J. Harrington, F. A. Smith, J. I. Bloch, D. M. Boyer, and K. H. Freeman. 2005. Transient floral change and rapid global warming at the Paleocene–Eocene boundary. *Science* 310:993–996.

Zachos, J. C., M. Pagani, L. Sloan, and E. Thomas. 2001. Trends, rhythms, and aberrations in global climate 65 Ma to present. *Science* 292:686–693.

Zachos, J. C., L. D. Stott, and K. C. Lohmann. 1994. Evolution of early Cenozoic marine temperatures. *Paleoceanography* 9:353–387.

PHYLOGENY, SYSTEMATICS, AND TAXONOMY

Bardeleben, C., R. L. Moore, and R. K. Wayne. 2005. A molecular phylogeny of the Canidae based on six nuclear loci. *Molecular Phylogenetics and Evolution* 37:815.

Clutton-Brock, J., G. B. Corbet, and M. Hills. 1976. A review of the family Canidae, with a classification by numerical methods. *Bulletin of the British Museum (Natural History), Zoology* 29:119–199.

Dahr, E. 1949. On the systematic position of *Phlaocyon leucosteus* Matthew and some related forms. *Arkiv för Zoologi* 41A:1–15.

Flynn, J. J., N. A. Neff, and R. H. Tedford. 1988. Phylogeny of the Carnivora. In M. J. Benton, ed., *The Phylogeny and Classification of the Tetrapods*, vol. 2, *Mammals*, 73–116. Oxford: Clarendon Press.

Geffen, E., A. Mercure, D. J. Girman, D. W. Macdonald, and R. K. Wayne. 1992. Phylogenetic relationships of the fox-like canids: Mitochondrial DNA restriction fragment, site, and cytochrome *b* sequence analyses. *Journal of Zoology* 228:27–39.

Grewal, S., P. J. Wilson, T. K. Kung, K. Shami, M. T. Theberge, J. B. Theberge, and B. N. White. 2004. A genetic assessment of the eastern wolf (*Canis lycaon*) in Algonquin Provincial Park. *Journal of Mammalogy* 85:625–632.

Hough, J. R. 1944. The auditory region in some Miocene carnivores. *Journal of Paleontology* 22:573–600.

Hough, J. R. 1948. The auditory region in some members of the Procyonidae, Canidae, and Ursidae. *Bulletin of the American Museum of Natural History* 92:73–118.

Hunt, R. M., Jr. 1974. The auditory bulla in Carnivora: An anatomical basis for reappraisal of carnivore evolution. *Journal of Morphology* 143:21–76.

Huxley, T. H. 1880. On the cranial and dental characters of the Canidae. *Proceedings of the Zoological Society of London* 16:238–288.

Lawrence, B., and W. H. Bossert. 1974. Relationships of North American *Canis* shown by a multiple character analysis of selected populations. In M. W. Fox, ed., *The Wild Canids: Their Systematics, Behavioral Ecology, and Evolution*, 73–86. New York: Van Nostrand Reinhold.

Ledje, C., and U. Arnason. 1996a. Phylogenetic analyses of complete cytochrome *b* genes of the order Carnivora with particular emphasis on the Caniformia. *Journal of Molecular Evolution* 42:135–144.

Ledje, C., and U. Arnason. 1996b. Phylogenetic relationships within caniform carnivores based on analyses of the mitochondrial 12S rRNA gene. *Journal of Molecular Evolution* 43:641–649.

Lyras, G. A., and A. A. E. Van der Geer. 2003. External brain anatomy in relation to the phylogeny of Canidae (Carnivora: Canidae). *Zoological Journal of the Linnean Society* 138:505–522.

Martin, L. D. 1989. Fossil history of the terrestrial Carnivora. In J. L. Gittleman, ed., *Carnivore Behavior, Ecology, and Evolution*, 1:536–568. Ithaca, N.Y.: Cornell University Press.

Matthew, W. D. 1930. The phylogeny of dogs. *Journal of Mammalogy* 11:117–138.

Mayr, E. 1940. Speciation phenomena in birds. *American Naturalist* 74:249–278.

Mayr, E. 1969. *Principles of Systematic Zoology*. New York: McGraw-Hill.

McKenna, M. C., and S. K. Bell. 1997. *Classification of Mammals Above the Species Level*. New York: Columbia University Press.

Nowak, R. M. 1979. North American Quaternary *Canis*. *Monograph of the Museum of Natural History, University of Kansas* 6:1–154.

Phillips, M., and V. G. Henry. 1992. Comments on red wolf taxonomy. *Conservation Biology* 6:596–599.

Segall, W. 1943. The auditory region of the arctoid carnivores. *Field Museum of Natural History, Zoological Series* 29:33–59.

Simpson, G. G. 1945. The principles of classification and a classification of mammals. *Bulletin of the American Museum of Natural History* 8:1–350.

Stains, H. J. 1974. Distribution and taxonomy of the Canidae. In M. W. Fox, ed., *The Wild Canids: Their Systematics, Behavioral Ecology, and Evolution*, 3–26. New York: Van Nostrand Reinhold.

Tedford, R. H. 1976. Relationship of pinnipeds to other carnivores (Mammalia). *Systematic Zoology* 25:363–374.

Tedford, R. H., B. E. Taylor, and X. Wang. 1995. Phylogeny of the Caninae (Carnivora: Canidae): The living taxa. *American Museum Novitates* 3146:1–37.

Turner, H. N., Jr. 1848. Observations relating to some of the foramina at the base of the skull in Mammalia, and on the classification of the order Carnivora. *Proceedings of the Zoological Society of London* 16:63–88.

Van Gelder, R. G. 1978. A review of canid classification. *American Museum Novitates* 2646:1–10.

Wang, X., and R. H. Tedford. 1994. Basicranial anatomy and phylogeny of primitive canids and closely related miacids (Carnivora: Mammalia). *American Museum Novitates* 3092:1–34.

Wang, X., R. H. Tedford, and B. E. Taylor. 1999. Phylogenetic systematics of the Borophaginae (Carnivora: Canidae). *Bulletin of the American Museum of Natural History* 243:1–391.

Wayne, R. K. 1993. Molecular evolution of the dog family. *Trends in Genetics* 9:218–224.

Wayne, R. K., R. E. Benveniste, D. N. Janczewski, and S. J. O'Brien. 1989. Molecular and biochemical evolution of the Carnivora. In J. L. Gittleman, ed., *Carnivore Behavior, Ecology, and Evolution*, 1:465–494. Ithaca, N.Y.: Cornell University Press.

Wayne, R. K., E. Geffen, D. J. Girman, K.-P. Koepfli, L. M. Lau, and C. R. Marshall. 1997. Molecular systematics of the Canidae. *Systematic Zoology* 46:622–653.

Wayne, R. K., and S. J. O'Brien. 1987. Allozyme divergence within the Canidae. *Systematic Zoology* 36:339–355.

Wozencraft, W. C. 1989. The phylogeny of the recent Carnivora. In J. L. Gittleman, ed., *Carnivore Behavior, Ecology, and Evolution*, 1:495–535. Ithaca, N.Y.: Cornell University Press.

Zrzavý, J., and V. Ricánková. 2004. Phylogeny of recent Canidae (Mammalia, Carnivora): Relative reliability and utility of morphological and molecular datasets. *Zoologica Scripta* 33:311–333.

Zunino, G. E., O. B. Vaccaro, M. Canevari, and A. L. Gardner. 1995. Taxonomy of the genus *Lyca-lopex* (Carnivora: Canidae) in Argentina. *Proceedings of the Biological Society of Washington* 108:729–747.

SEXUAL DIMORPHISM

Gingerich, P. D., and D. A. Winkler. 1979. Patterns of variation and correlation in the dentition of the red fox, *Vulpes vuples*. *Journal of Mammalogy* 60:691–704.

Gittleman, J. L., and B. Van Valkenburgh. 1997. Sexual dimorphism in the canines and skulls of carnivores: Effects of size, phylogeny, and behavioural ecology. *Journal of Zoology* 242:97–117.

Jolicoeur, P. 1974. Sexual dimorphism and geographical distance as factors of skull variation in the wolf *Canis lupus* L. In M. W. Fox, ed., *The Wild Canids: Their Systematics, Behavioral Ecology, and Evolution*, 54–61. New York: Van Nostrand Reinhold.

Kolenosky, G. B., and R. O. Standfield. 1974. Morphological and ecological variation among gray wolves (*Canis lupus*) of Ontario, Canada. In M. W. Fox, ed., *The Wild Canids: Their Systematics, Behavioral Ecology, and Evolution*, 62–72. New York: Van Nostrand Reinhold.

Prestrud, P., and K. Nilssen. 1995. Growth, size, and sexual dimorphism in arctic foxes. *Journal of Mammalogy* 76:522–530.

Regodon, S., A. Franco, J. M. Garin, A. Robina, and Y. Lignereux. 1991. Computerized tomographic determination of the cranial volume of the dog applied to racial and sexual differentiation. *Acta Anatomica* 142:347–350.

ZOOGEOGRAPHY

Berta, A. 1987. Origin, diversification, and zoogeography of the South American Canidae. In B. D. Patterson and R. M. Timm, eds., *Studies in Neotropical Mammalogy: Essays in Honor of Philip Hershkovitz*, 455–471. Fieldiana Zoology, n.s., 39. Chicago: Field Museum of Natural History.

Brunet, M., F. Guy, D. Pilbeam, D. E. Lieberman, A. Likius, H. T. Mackaye, M. S. Ponce de Leon, C. P. E. Zollikofer, and P. Vignaud. 2005. New material of the earliest hominid from the upper Miocene of Chad. *Nature* 434:752–755.

Hunt, R. M., Jr. 1996. Biogeography of the order Carnivora. In J. L. Gittleman, ed., *Carnivore Behavior, Ecology, and Evolution*, 2:485–541. Ithaca, N.Y.: Cornell University Press.

Johnson, W. E., T. K. Fuller, and W. L. Franklin. 1989. Sympatry in canids: A review and assessment. In J. L. Gittleman, ed., *Carnivore Behavior, Ecology, and Evolution*, 1:189–217. Ithaca, N.Y.: Cornell University Press.

Qiu, Z. 2003. Dispersals of Neogene carnivorans between Asia and North America. *Bulletin of the American Museum of Natural History* 279:18–31.

Stains, H. J. 1974. Distribution and taxonomy of the Canidae. In M. W. Fox, ed., *The Wild Canids: Their Systematics, Behavioral Ecology, and Evolution*, 3–26. New York: Van Nostrand Reinhold.

Vilà, C., I. R. Amorim, J. A. Leonard, D. Posada, J. Castroviejo, F. Petrucci-Fonseca, K. A. Crandall, H. Ellegren, and R. K. Wayne. 1999. Mitochondrial DNA phylogeography and population history of the grey wolf *Canis lupus*. *Molecular Ecology* 8:2089–2103.

INDEX

body size (*continued*)
hesperocyonines, 105, 120, *121*; of hyaenids, 105–106; and hunting, 103–106; in Ice Age, 134; of mammals of late Eocene, 120, *121*; of mammals of late Miocene, *128–129, 131*; of mammals of late Oligocene, *124–125*; of mammals of Pleistocene, 134–135; and sexual dimorphism, 107–108

bone cracking, 112, 130, 132; teeth for, 12, *13*, 24, 25, 33, 36, 46, 74, *82*, 83, 88–89, 130

bone crushing. *See* bone cracking

Borhyaena, 19

Borhyaena tuberata, *18, 19*

Borhyaenidae, 7, 17–19

Borophaginae, 31, 33–49, 53; body size of, 36, 105, 124, *131*; brain of, 110; coevolution of, with prey, 130; endemism of, 140; evolution of, viii, 24, 35, 49, 51, 122, *126*, 127, 128, 130, 131, 132, 137, 178; functional morphology of, 80, 112; geographic dispersal of, 142, 143; intercontinental migrations by, *141*; skull and neck of, *100*; species of, 170–171; teeth of, 33–35, 73, 89, 111, 124

Borophagini, 171–172

Borophagus, 33, 36, 46; body size of, 36, 105, 124; brain of, 110; evolution of, 34, 36, 46, 130, 132, 178; geographic dispersion of, 143; jaw of, 83; as scavenger vs. hunter, 112; teeth of, 46, *82*, 89

Borophagus diversidens, 36, 49, 52, 83, 132, 133, 172, *plate 7*

Borophagus dudleyi, 172

Borophagus hilli, 172

Borophagus littoralis, 36, 172

Borophagus orc, 36, 172

Borophagus parvus, 172

Borophagus pugnator, 172

Borophagus secundus, *50*, 89, 91, 112, *113*, 172, *plate 5*

brown hyena. See *Parahyaena brunnea*

browsers, 128

buffalo, 134

bush dog. See *Speothos venaticus*

Caedocyon, 26, 178

Caedocyon tedfordi, 26, 170

camelids, 140

canid anatomy, 69–101; body size, 103–106, *106*, 107, 135; brain, 99, 101, 110, 165; claws, 94–96; dome of head and frontal sinus, 83; ear bones, 83–84, *85*; jaw, 3, 4, 75, 80, 83, 158; limbs, 89, 91–94, 97, 114; maxilloturbinates, 86–87, 88; neck, 96–99, *100*; paedomorphic features, 158, 159, 168; phalanges, 92–94, *95*; rostrum, 79–82, 158; skull, 3, 4, 69, 75, 78, *85*, 158; standing posture, 89, 91–94; turbinates, 85–88. *See also* teeth

canid evolution, viii, 22; convergent, with hyaenids, 116, 142; in Eocene, 119–120, *126, 141*; in middle Miocene, 125–131, *126*, 137, *141*, 142, 143, 145, 148, 150; in Oligocene, 120, 122–125, *126, 141*; in Paleocene, 117–119; and physical environment, 117–138; in Pleistocene, 134–137, *141*, 149, 150; in Pliocene, *126*, 131–134, 137, *141*, 145, 146, 147, 148, 150; of proto-canids, 119; timeline of, *118*

Canidae, viii, 1, 3, 12, 20, 49, 70; domestication of, 135, 154–166; endemism of, 139–143; in Eurasia, 12, 52, 58, 61, 132, 135, *141*, 144, 148; fossil record of, viii, 20–22, *22*, 23, 69, 120, 122, 144–145, 147, 154–157, 158, 161–162; geographic dispersal of, 143–151; hunting behavior and social activity of, 103–116; intercontinental migrations by, *141*, 144, 148; mating systems of, 107; phylogeny of, 4; sexual dimorphism of, 107–108; species of, 23; subfamilies of, 23–67; zoogeography, 139–151. *See also* canid anatomy; canid evolution; Caninae; *Canis*

Caninae, viii, 24, 49, 51–65, 97; evolution of, 53, 122, *126*, 131, 132, 137, 179; geographic dispersal of, 143–151; head of, 80, 97; intercontinental migrations by, *141*; species of, 172–175

Canis, 60–65, 174; body size of, 135; evolution of, 52, 58–60, 148, 179; geographic dispersal of, 158–160; morphology of, 110; sexual dimorphism of, 107

Canis adustus, 174

Canis africanus, 174

Canis antonii, 174

Canis armbrusteri, 52, 53, 60, 149, 174

Printed in the USA
CPSIA information can be obtained
at www.ICGtesting.com
JSHW051458221024
72172JS00011B/99